# Concrete and
# Statistics

# Concrete and Statistics

by

J. D. McINTOSH
M.Sc., A.M.I.C.E.

CR BOOKS
A Maclaren Company
LONDON

85334 038 2
CONCRETE AND STATISTICS
BY J. D. MCINTOSH
FIRST PUBLISHED 1963 IN
THE CONCRETE LIBRARY
BY C. R. BOOKS LTD.
THIS 2ND PRINTING
FIRST PUBLISHED 1968 BY
C. R. BOOKS
THE MACLAREN GROUP OF COMPANIES
7 GRAPE STREET LONDON WC2

PRINTED IN GREAT BRITAIN BY
LOWE AND BRYDONE (PRINTERS) LTD. LONDON

TO

MY FAMILY

# Preface

'Statistics' is a word that may have different associations for different people. For some it may carry the meaning implied in the comment 'there are lies, damned lies and statistics'; for others it may always be preceded by the word 'vital'; or it may conjure up tedious form-filling so that someone can produce statistics. In the context of this book statistics is a branch of mathematics; it can be highly analytical, but only the simpler concepts are used here.

The author makes no claim to be a mathematician, let alone a statistician. But, as a concrete technologist, he has gained a more realistic view of his subject by an examination of several problems from a statistical point of view. The book is an attempt to introduce these statistical concepts to others concerned with the same subject, and to stimulate thinking on these lines. Much of it deals with what are really simple, common-sense criticisms of current practices, brought home more forcibly by an appreciation of statistical method. On the other hand, guidance is given on the drafting and enforcing of specifications for cement, aggregates and concrete, related to statistical aspects of structural safety, and on the procedures for controlling the quality of these materials so that they will comply with relevant specifications. The purpose of using statistics is to obtain the maximum amount of accurate information from the limited number of test results which can normally be obtained in practice. Brief reference is made to simple investigations, but no attempt is made to cater for the professional research worker who may require a much more detailed knowledge of statistical mathematics to carry out his experiments with maximum efficiency. References to books and articles are given for the benefit of those who wish to pursue the subject further.

Many of us do not have the background which an appreciation of statistical method provides because our normal educational system generally teaches us mathematics in terms of numbers which are absolute and infinitely precise and related only to idealised situations. Even in applied mathematics we get little guidance on how to apply the mathematical procedures to everyday problems as they actually exist rather than as simplified examples. Times are changing, and statistics may be offered as one of the pure mathematics papers in the University of London General Certificate of Education examination at Ordinary level, but it does not form part of the syllabus at Advanced or Scholarship level.

In practice, numbers are usually associated with measurements and these are subject to errors due to lack of precision in the equipment used to obtain them, due to human weakness in the operator making the tests and due to differences in the samples taken for examination, both from one another and from the mass of material being inspected. Statistical procedures are used to study these errors and variations, especially if their occurrence is purely random, with a view to obtaining a more realistic numerical interpretation of data.

While statistical calculations are a means of extracting the maximum amount of information from a set of test results, there are nevertheless certain limitations to what can be done: it is these limitations that help most in providing a sense of proportion to our interpretation of results. For example, statistical methods show that there is never an absolute answer to a question which may be asked in making an experimental investigation: the answer is always in terms of probability, based on the evidence put to the test of the statistical procedures. Obviously, too, statistics cannot produce more information than is available in the data to be examined by these methods, but a knowledge of the subject helps us to plan our testing so that the maximum amount of information can be gained from a minimum of effort.

For this purpose general experience of the subject under examination is also essential to make the best use of statistics. This book has been written primarily for those concerned with making concrete, and the experience to which statistical procedures are related should be well within the range of most people connected with the concrete industry. There is no need for all who are concerned with the quality of concrete to become expert statisticians—many may never be called upon to make a statistical calculation—but there is a pressing need for them to have a background knowledge of statistical concepts which will enable them to bring a more balanced judgment to questions of specification and control.

Most of the discussion bears ultimately on the drafting and implementing of specifications; control testing and preliminary investigations are both means by which more economical compliance with a specification is achieved. A specification is a legal document, forming part of the contractural agreement between two parties, which states the basis for accepting or rejecting goods according to their quality when sold by the one to the other. Its sole purpose is to make clear the intention of both parties to each other in this respect, and therefore it should be devoid of ambiguity and 'wishful thinking' and should be capable of being enforced.

Writers of specifications would do well to read Shakespeare's 'The Merchant of Venice', Act IV, Sc. i. There the argument lies not in the question of whether one party has failed in his bargain with the other, but in the understanding of the enforcement of the forfeit to be paid. This is true of many concrete specifications which read as though their authors took the same attitude as Shylock:

> 'The pound of flesh, which I demand of him,
> Is dearly bought; 'tis mine and I will have it'.

They have not thought of the implications:

> 'The bond doth give thee here no jot of blood;
> The words expressly are "a pound of flesh".'

and:

> ' . . . if thou cut'st more
> Or less than a just pound, be it but so much
> As makes it light or heavy in the substance,
> Or the division of the twentieth part
> Of one poor scruple, nay, if the scale do turn
> But in the estimation of a hair . . . '

It is to be hoped that this study of statistical ideas will introduce greater realism into concrete specifications so that

> ' . . . earthly power doth then show likest God's
> When mercy seasons justice'.

# Acknowledgments

The book summarises the author's experience of the subject gained through his employment with the Cement and Concrete Association: he wishes to acknowledge the excellent facilities available for the discussion and criticism of developing ideas and for the accumulation of factual information. He recognises the part played by his many colleagues who have contributed to the publication of the book, either wittingly, by making test results available, or unwittingly, in general conversation.

Especial recognition, however, is due to Mr. H. C. Erntroy, M.Sc., A.M.I.C.E., who has discussed the details and implications of most statistical problems associated with concrete technology, to Mr. J. D. Dewar, B.Sc.(Eng.), who has helped with the reading of the manuscript, and to Prof. A. Markestad for his contribution of Scandinavian experience. The production of the book has been helped greatly by Mrs. Davy who undertook most of the computational work, by Mrs. Barnes who typed successive drafts, and by Mrs. Palmer who prepared the tracings for the figures.

# Contents

# List of Tables

# List of Figures

# CHAPTER ONE

# Numbers, Measurement and Variation

## PURE NUMBERS

Our present educational system ensures that we shall all be given some instruction in mathematics, but mathematics as it has been taught is an exact science. This means that we are able, in varying degrees, to manipulate pure numbers. However, when we come to grips with the world around us, most of us soon find that our mathematics is rarely pure or exact.

Even if the subject we are studying involves a precise number of entities (which is rare), it is not always easy for us to count them and arrive at the same total; for example, women often need the help of someone else in counting the stitches in a row of knitting before they are sure they have the correct answer. Then, when we manipulate our pure numbers we do not always arrive at the same answer because of our human weaknesses. Again, as an example, when adding up a column of figures, our minds can easily be distracted and we can obtain different answers at different attempts; even with a calculating machine we can make the mistake of pressing the wrong keys.

However, in both these examples there is a correct and precise answer to the question we are asking and we can go on checking and rechecking until we 'know' we are right because we are conscious of having made no mistake. With sufficient effort the error due to the 'operator' can be eliminated.

## PRECISION OF MEASUREMENTS

When we come to try to measure some property of material, such as its length, its weight, or its strength, we have first of all to define that property and then find a means of measuring it. Quite often, and especially in recognised standards, the definition is given in terms of a detailed test procedure. Where the property is to be measured to check compliance with the requirements of a legal document, this is preferable to a descriptive definition, which could be construed in different ways, as Shylock found to his cost over the meaning of the word 'flesh'. Even if he had not made this mistake, he would still have failed in his case because he fell victim to the idea that it was possible to measure precisely 'a pound of flesh', and he gave himself no tolerance to allow for the smallest of errors.

No matter how clever the authors of a test procedure may be, they can never foresee all the factors that will affect the result, nor would it be economical to specify them; there is therefore no such thing as a precise measurement of that property. If we repeat the test on the same sample a number of times under varying conditions, but complying strictly with the specified test procedure, we will have a series of different values of the property, all of which have been obtained from accurate measurements according to our definition. This is true even if we are dealing with testing of the highest precision. In practice we can only afford to test samples of the material—and not the whole lot—and one sample is unlikely to be just the same as another. We are no longer dealing with an exact science.

To be fair to our school teachers, they must have foreseen this situation because they have taught us how to calculate the average, or the arithmetical mean, of a set of numbers. So, naturally, we feel that we must determine the average of our series of test results and take this as a 'better' value of the property than any of the individual values making up the average. In fact, the average is not necessarily any 'better' than any of the individual results which have been obtained according to the test procedure; but it is subject to less variation and is therefore more reproducible. If we were to do other series of similar tests and calculate the average values from the new series we would probably find that we still had different answers, but the differences in the averages would probably be less than the differences in the individuals in any one set. All the average values cannot be the 'true' value—they can only be estimates of the 'true' value—if by that we mean the ultimate average value, and obviously it would be necessary to make an infinite number of tests to obtain what might be referred to as absolute precision in our determination of this value.

## ACCURACY NEEDED FOR SPECIFICATIONS

It follows, therefore, that the result obtained from any limited series of tests is only an approximation to the precise answer which would be obtained if all the unknown factors affecting the result operated at their average value. In practice we want to know whether the result approximates closely enough for our purpose; sometimes a rough answer may be all that is required, but for checking for compliance with a specification we need a testing technique which gives us results which are as closely reproducible as is economically justified so that there will be little chance of ambiguity. This is required not only to save pointless argument when different people test the same material and obtain different results, even though they are all obtained from the correct procedure, but also to ensure that satisfactory material is not needlessly rejected or that unsatisfactory material is unjustifiably accepted because of variations resulting from our testing method.

When a specification states limits for a particular property, allowance should have been made for the probable magnitude of the variations associated with the standard procedure for measuring that property. The greater are these variations in relation to the true differences likely to be encountered, the harder it is to fix satisfactory limits because producers and users both want to be given the benefit of any doubt arising from the presence of these variations. For example, the variations associated with testing single cubes sampled from individual batches of concrete are such that about 19 out of 20 of the observed values would be within $\pm$ 400 lb/in$^2$ of the 'true' value for the batch when the minimum strength of site-made concrete is about 3,000 lb/in$^2$. The supervising engineer who wants the minimum strength of the batch to be 3,000 lb/in$^2$ would be pleased to have the test limit based on a single specimen set at 3,400 lb/in$^2$ to guard against having more than occasional batches of material supplied with an actual strength of less than 3,000 lb/in$^2$ because of a testing variation greater than 400 lb/in$^2$ operating in that direction. On the other hand, the contractor or ready-mixed concrete supplier would feel the limit should be only 2,600 lb/in$^2$ to prevent more than occasional batches of concrete being condemned just because a test result carried with it a variation of the same amount, but in the opposite direction.

## ERRORS

The variation in the results we have obtained from our measurements of compressive strength have not been due to basic differences in the materials tested, but are due to the shortcomings of the testing and sampling procedures employed; they are referred to as the 'testing error' and 'sampling error'. The testing error is caused by changes in factors which are not controlled in the specified test procedure, but nevertheless affect the result; they are generally independent of the sampling procedure, but may depend on the magnitude of the property of the material being tested.

The sampling error depends, as would be expected, on the particular sampling procedure adopted, but it also depends considerably on the variability within the consignment of material being sampled. The sampling procedure may be modified according to the purpose for which the tests are being made and can be arranged either to accentuate or to minimise the inherent variability in the consignment; the term 'consignment' is used to denote the whole of the material being examined, including those parts which are left after samples have been taken for testing. If we are interested in the 'true' differences between the average properties of one consignment and those of another, as, for example, between two batches of concrete, the sampling procedure should be arranged to minimise the inherent variability within the consignment or batch, otherwise the sampling error, coupled with the testing error, may well mask the true difference between the different consignments; under these conditions, the sampling procedure accomplishes mechanically what averaging does mathematically, but the cost is usually much less. Obviously, the impression of the quality of a material we obtain from test results depends on the relationship of the sample of material examined to the whole consignment, as given by the sampling procedure.

## SAMPLES

In general, when we require a sample of a material for test we cannot be satisfied with just any part of that material; to be a sample the part must in some way represent the whole consignment, and the result of the tests can give no more information about a consignment than the relationship of the sample to the consignment permits.

Where testing is being carried out for specification purposes the fate of a whole consignment may easily depend on the care with which the samples have been taken. It is surprising, therefore, to find so often that the job of sampling is left to the lowest grade of labour available and that no instruction whatever is given in the art of sampling; yet those responsible for having the test made would probably object strongly if the person employed in conducting the test were not someone having adequate experience and qualifications. An appreciation of the variations involved in sampling and testing helps one to attach proper significance and importance to the art of sampling.

Occasionally we may be satisfied with what we may call a 'typical' sample. By this we do not necessarily mean a sample whose properties are the arithmetical average for the consignment, but a sample which should not be extreme in its properties, perhaps due to it being taken under some conditions which are not normal for the material: the sampler should have sufficient experience to know what is normal. But for most specification testing the procedure by which samples are to be obtained is carefully described so that the relationship of the sample to the whole consignment is predetermined. Usually we attempt to get a sample which represents the average for the consignment.

However, no matter how carefully the sampling procedure is described, the property of the sample will generally be somewhat different from the average for the consignment as a whole and the result we obtain from our tests on the sample is only an estimate of the property in which we are interested. We can improve the accuracy of our estimate by either increasing the size of the sample or increasing the number of samples we take for test until, in the end, we might test the whole of the consignment. But such a procedure would not be of economical value—especially when the property is measured by a destructive form of test, as are most of the tests associated with concrete technology.

The size of the sample is governed largely by the quantity of material required for the standardised test procedure and there is little opportunity for increasing the amount used in the test. If, however, one can improve the uniformity of a material by subsequent laboratory treatment, such as more thorough mixing, the same effect can be obtained by sampling a larger quantity of material, mixing it thoroughly, taking the same amount as before as a sample of the initial sample and testing it. This procedure is adopted for the testing of aggregates, for example, where the large initial sample is obtained as a large number of separate 'increments' which are mixed together and the appropriate quantity for test is obtained by 'quartering' or by the use of a riffle box;[1] similarly, in sampling concrete, several small samples of concrete should be mixed together to form an initial sample and then the quantity required for a test is extracted.[2] By this means we are able to obtain for a given number of tests a more accurate estimate of the average quality of the consignment than would have been possible if the same number of tests had been made on separate small samples.

If we are also interested in the variation in the property occurring within a consignment, we should make a large number of tests, each on separate samples taken from the consignment, and the number of samples is more important than the size. If the information is required about one particular consignment, the several samples will obviously have to be taken at the same time. But quite often we are interested in the variation of a material from time to time; to obtain a good assessment of this variation in consignments it is better to increase the frequency of the taking and testing of separate samples than the number of samples taken at any one time.

Other forms of sampling procedure may be adopted if we are also interested in measuring the sampling and testing errors. No further distinction will be made between sampling and testing errors and the term 'testing error' will be used to include both.

## ROUNDING OFF

If our test procedure is one that requires a mathematical calculation to derive the result, as, for example, the measurement of compressive strength of concrete, which requires the division of the load at which a specimen fails by the cross-sectional area, our love of mathematics as a pure science leads many of us to work out and state the calculated value to too many significant figures. But when we think about it, we can use our judgment, based on past experience, to guide us in the accuracy with which we record our result. For example, it should be obvious, after a little experience with concrete, that it is pointless to report individual values of compressive strength as closely as the nearest 1 lb/in² because of the errors which lead to differences in results, for similar specimens, of the order of some hundreds of lb/in², as already pointed out earlier in this chapter. On the other hand, we would know that it would be stupid not to report our results to values somewhat closer than 1,000 lb/in². With experience we would probably say that it would be appropriate to report our results to somewhere between 10 and 100 lb/in², but the actual value we would choose should depend, among other things, on the details of the sampling and testing procedures; the accuracy of reporting average values also depends on the number of repeat tests included. The method of rounding off appropriate to a particular standard test procedure should be given in the specification: *e.g.* B.S. 1881, 'Methods of Testing Concrete',[2] requires that the compressive strength of individual test cubes be quoted to the nearest 50 lb/in².

Statistical mathematical procedures are available to guide us more closely than can our judgment in this matter. As a guide, values should be rounded off to a convenient interval less than half the standard deviation of the values;[3] the standard deviation is defined in the next section. The important point for the present is that our assessment of the accuracy of the result (whether derived from statistics or judgment) is based on the variation or scatter in the results that we obtain in practice when testing a nominally uniform sample of material according to the standardised procedure.

## MEASURES OF VARIATION

The variation in a set of test results is the main subject of our statistical thinking. We have probably been taught at school that we can use the ' range '—the difference between the highest and the lowest number in a set—as a measure of the variability of the numbers. But we soon realise that the range is likely to vary with the number of results in the set, and that it is dependent on the two extreme values only, which may be far from representative, especially if we have a large number of values in the set.

With a little experience we learn to assess intuitively the importance of the numbers lying between the extreme values, and we soon acquire a sense of what is good or poor reproducibility based on the whole set of results. This is particularly necessary if we are trying to compare the value of some property before and after a change has been made which does not produce a very large difference. We need to know whether the difference between the average values, before and after the change was made, is appreciably greater than the difference found within either series of tests. We may decide that there is a good chance or only a poor chance that the change has produced what we feel to be a significant difference. This problem is discussed in relation to trial mixes in Chapter 6 and Appendix 1.

The 'mean deviation' of a set of numbers is a better measure of the variability of those numbers than is the range because all the numbers in the set are used in its calculation. It is obtained by first determining the average of the set and then subtracting the average from each individual number in turn; ignoring the positive and negative signs of these differences (or deviations), the mean value is calculated as the mean deviation. Unfortunately, the mean deviation is not particularly suitable for further calculation of characteristics of the variability of the set of numbers.

The 'standard deviation', however, is a much more convenient measure of variability; it is based on the squares of the deviations, but is not easily described in words. The standard deviation is defined numerically as:

$$s = \sqrt{\frac{\Sigma (x - \bar{x})^2}{n}}$$

where  s = the standard deviation of the set of numbers,

$\bar{x}$ = the average of the set of numbers,

x = any value in the set of numbers,

n = the number of values in the set.

($\Sigma$ indicates that all the values of $(x - \bar{x})^2$ have to be added together.) Like the range and mean deviation, the standard deviation increases with increasing variability and is expressed in the same units as the numbers.

However, we are less often interested in the standard deviation of the actual numbers in the set, obtained from the tests on the sample, than we are in the variation of the property of the consignment from which the sample was drawn; for example, we are more likely to be interested in an estimation of the variation in the compressive strength of a whole day's concreting than we are in the actual variation of the results for the relatively few batches which have been sampled and tested. One could take the value of 's' for the sample for this purpose, but a better estimate is obtained from the formula:

$$\sigma = \sqrt{\frac{\Sigma (x - \bar{x})^2}{n - 1}}$$

where $\sigma$ = the estimate of the standard deviation for the whole consignment.

The difference in the two formulae arises because $\bar{x}$ is only an estimate of the average value for the whole consignment and the inaccuracy brings with it a bias to the calculation of the standard deviation for the consignment. The sum of the squares of the deviations of individual values from the true average for the consignment would be greater than $\Sigma(x - \bar{x})^2$, by an amount depending on the number of values in the set, and this can be compensated for by substituting 'n − 1' for 'n' in the denominator of the expression. Throughout the remainder of the book, unless otherwise stated, the term 'standard deviation' refers to the estimate for the consignment and not the defined value for the sample; the appropriate formula, therefore, is that for '$\sigma$' and not 's'.

The 'coefficient of variation' is a non-dimensional measure of variation obtained by dividing the standard deviation by the average, thus:

$$v = \frac{\sigma}{\bar{x}} \times 100$$

where $v$ = the estimate of the coefficient of variation of the consignment, expressed as a percentage.

## USE OF THE STANDARD DEVIATION

The standard deviation has the valuable property that the proportion of all the results falling within or outside certain limits can be related to it when the results from which it has been calculated, and all the results with which we are concerned, occur at random. This assumption can generally be made in concrete work without serious loss of accuracy as long as a reasonable attempt is made to sample at random. If limits are set at levels equal to the average minus the standard deviation and the average plus the standard deviation ($\bar{x} \pm \sigma$), about two-thirds of the results would be within these limits. If the standard deviation were known precisely for an extremely large number of results, the accuracy of quoting the proportion of results could be correspondingly increased because it is obtained by calculation from a formula for the ideal situation; the value would then be 0·6827.

This use of the standard deviation may be illustrated pictorially by a diagram known as a histogram (see fig. 1). The horizontal scale is in the same units as the property whose standard deviation is under examination; for example, in studying the variability of compressive strength, the scale would be graduated in $lb/in^2$. This scale is divided into a number of equal intervals spanning the range of the results, and at the centre of each interval we plot vertically the number of results falling within that interval. With a random distribution there is a tendency for the greatest numbers of results per interval to occur in the intervals nearest the average value and for the numbers to tail off the further the interval from the average. As the number of results available increases, the interval could be made smaller and the envelope of the

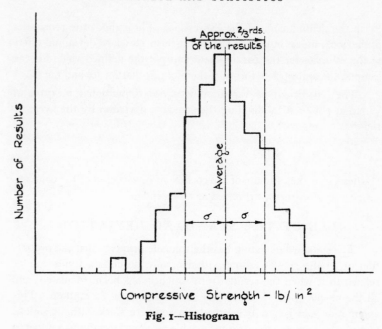

Fig. 1—Histogram

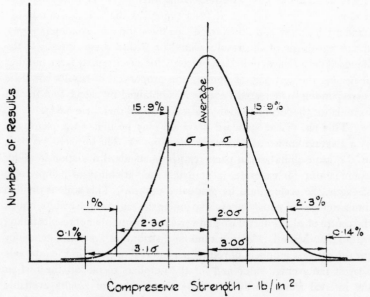

Fig. 2—Normal probability curve

histogram would tend to become smoother, until in the idealised situation, representing a purely random occurrence, the envelope would become a smooth curve, known as the normal probability curve, corresponding to the formula given in Appendix 1.

In dealing with the histogram or the probability curve we are not usually interested in the ordinates (which are termed the 'population density'), as they have little practical significance: the important feature is the area under the curve to either side of an ordinate, as shown in fig. 2. The area under the whole curve represents the total number of results with which we are concerned, and the areas below the curve bounded by ordinates on either side represent proportions of the total number of results. The vertical scale for the normal probability curve is usually arranged so that the area under the curve is unity.

It will be seen that the normal probability curve is symmetrical about an ordinate and that there is the greatest chance of results occurring at this value. This is the arithmetical mean value and the equal areas below the curve on either side indicate that there will be as many results above the mean as there are below it. If ordinates are drawn on either side of the mean ordinate at distances equal to the standard deviation the proportion of results falling between the limits of $\bar{x} - \sigma$ and $\bar{x} + \sigma$ is indicated by the area below the curve between these two ordinates.

The characteristics of the curve are fixed by the average value and the standard deviation. The spread of the curve along the horizontal scale is governed by the standard deviation—higher values leading to a wider distribution or greater spread; thus the shape of the curve is established. The position of the curve along the scale is fixed by the average value.

Because about two-thirds of the values lie within $\bar{x} \pm \sigma$, it follows that about one-third must lie outside this range. Furthermore, as the variations occur at random, they must be just as likely to occur below the average as above, so that we can expect about one-sixth of the results to occur below $\bar{x} - \sigma$ and the other one-sixth to occur above $\bar{x} + \sigma$. As we are usually concerned with limiting values for properties, such as a minimum compressive strength or a maximum slump, this is a useful way of employing the standard deviation.

We can choose proportions of results other than one-sixth and fix the corresponding limits below or above which the proportion of results can be expected to fall: the limits would be $\bar{x} - k\sigma$ or $\bar{x} + k\sigma$

C

where k has the values given in fig. 3 appropriate to the chosen proportion of results. For example, only 1 out of 100 results would be
expected to fall below $\bar{x} - 2 \cdot 33\sigma$. This is also illustrated in fig. 2.
It is possible to achieve a probability of, say, 1 in 100 results falling
below a particular value in many ways by choosing corresponding
values of $\bar{x}$ and $\sigma$ to satisfy the condition: the lower the value of the
standard deviation, $\sigma$, for a set of results, the lower will be the average,
$\bar{x}$, needed to meet the requirement.

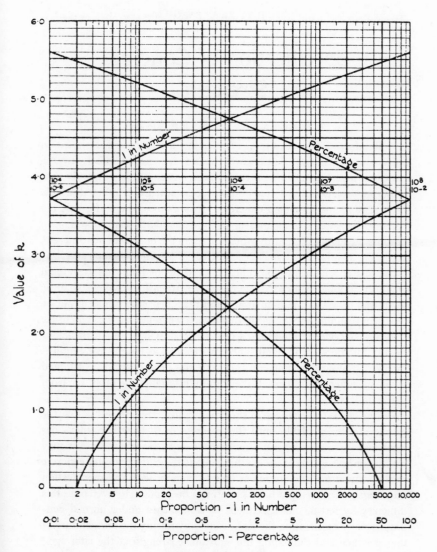

Fig. 3—Values of k

# CHAPTER TWO

# Specification of Concrete Strength

## COMPRESSIVE STRENGTH

Reinforced and prestressed concrete structures are designed in such a way that the compressive stresses are carried mostly by the concrete itself; even where compressive strength is relatively unimportant structurally it can be used as a measure of other desirable properties of the hardened concrete. It is natural, therefore, that the compressive strength, as measured by a standard test, should have become the main criterion by which to assess the quality of the concrete, especially as it is easy to make the test cubes or cylinders on a site.

Most specifications for concrete used in medium and large civil engineering jobs require that compressive strength specimens shall be made and tested at a particular age and that their strength shall not be less than a minimum value. By this means we hope that we can ensure that the concrete, produced as a raw material, will be of a satisfactory quality for the work in hand. This works well as long as the requirements of the specification are met, but the procedure to be adopted when specimens fail to reach the minimum strength is often far from being understood.

The chief disadvantage of the compressive strength test is that the results are not available until the concrete in the test specimens has hardened for the appropriate number of days, by which time it may well be impracticable to do anything about the corresponding concrete in the structure if the specimen fails to comply with the specification requirements. Attempts have sometimes been made to overcome this difficulty by testing the specimens at earlier ages under normal conditions or by accelerating the test, usually by raising the temperature of the specimen and estimating the equivalent strength at the later age.[4] If there were excellent correlation between the result obtained from any accelerated test and the result from the corresponding test under normal conditions, more effective action could be taken to deal with unsatisfactory concrete than is possible when the results are not available for at least 7 days and probably 28 days. Unfortunately, as yet, it seems that no accelerated test has been devised which gives a sufficiently close correlation for all conditions of materials and mix proportions for the result of a single accelerated test to be sufficient evidence of the failure of the concrete it represents to comply with the requirements of the standard, expressed in terms of test results on specimens cured at normal temperatures. This being so, it would be unwise to use the accelerated methods for work where only small numbers of test specimens would be made and concrete would be rejected on the basis of any one result.

It seems, therefore, that the engineer specifying the quality of concrete is driven back to requiring certain strengths from specimens made according to a standard procedure, cured at a standard temperature in the normal working range and tested by a standard method at an age approaching that at which he is interested in the strength. A procedure for making, curing and testing cubes has been laid down for him already in B.S. 1881 : 1952,[2] and other standards are applicable elsewhere, but he still has the choice of the age of test. The method of interpreting the results so that the requirements of the specification can be understood without ambiguity and can be enforced reasonably is less well defined.

## INTERPRETATION OF MINIMUM STRENGTH

If we were to examine the results of compressive strength tests on a large number of specimens taken at random from a continuous run of production of concrete of nominally uniform quality we would expect to find that they would tend to follow the pattern shown in fig. 1. The corresponding theoretical probability curve, like that in fig. 2, extends indefinitely in both directions, so that in this sense there is no such thing as a minimum or a maximum value, *i.e.* there is always a chance that a result will occur with a value as low or as high as the circumstances of the case permit. Indeed, we have already seen when considering the different measures of variation that the range of a set of numbers is likely to increase as the number of values in the set increases. Thus, from the statistical point of view, we can see that the term 'minimum strength' can only have any significance if it is related to a probability that a certain proportion of the results will fall below that value. Furthermore, if the minimum value is defined in this way its relationship to the average value is known in terms of the standard deviation.

It has long been appreciated that the compressive strength of concrete made on site is subject to wide variations and that the average strength to be aimed for must be appreciably higher than the minimum if the quality of the concrete is to comply with the requirements of the specification. For some time use has been made of the statistical approach to the relationship between the specified minimum strength and the required average strength to help to establish reasonable and economical methods of producing concrete complying with the specification. This aspect of mix design may be illustrated as follows.

If, from previous experience, we could expect the variation in the compressive strength to be represented by a certain standard deviation, we could choose an average strength for which the mix would be designed, which would carry with it a predetermined chance of results falling below a specified minimum strength. For example, if the expected standard deviation is 850 lb/in², the specified minimum strength is 3,000 lb/in² and we expect only 1 in 100 results below the minimum, the mix would be designed for an average strength of:

$$3,000 + 2 \cdot 33 \times 850 \text{ lb/in}^2$$
$$= 5,000 \text{ lb/in}^2$$

After a concreting job has got under way, statistics can be used to help to forecast whether the quality is likely to comply with the requirements of the specification. For example, if we are making concrete which has to comply with a requirement for a certain minimum strength, and we obtain a number of values of compressive strength from samples taken at random from the concrete which has been made so far, we can estimate the chances that a result will fall below the specified minimum value if, indeed, none has already done so. We would calculate the values of $\bar{x}$ and $\sigma$ from the results, subtract the specified minimum from $\bar{x}$, and divide this difference by $\sigma$, thus giving a value of k. The corresponding chance of failure would be found from fig. 3.

Interest in the use of statistics to help with the design of mixes and control of concrete quality on the site appeared to assume practical significance during the late 1940s.[5-8] Its importance can be judged by the fact that many of the papers presented to the Symposium on Mix Design and Quality Control of Concrete held in 1954[9-16] made some reference to these techniques. Most of the papers and discussions, as recorded in the proceedings of that symposium, are still relevant and well worth reading. A brief synopsis of papers relevant to this subject is given in Appendix 3.

## MINIMUM STRENGTH SPECIFICATION

There is no simple specifying procedure suitable for all types of concrete work and the scheme to be adopted will depend on the type of structure, the importance of the work, the margin of safety in the design and the number of specimens likely to be made. For example, if the concrete is being used for precast members, there would be relatively little complication in rejecting units which were made of concrete giving sub-standard cube results before they were built into the structure, whereas it may be extremely difficult to cut out and replace concrete in a foundation block if, in the days from casting to testing the specimens, a considerable amount of superstructure had been erected. The need for drastic action in the event of specimens failing to meet the requirements would be much greater if failure of the structure could lead to considerable loss of life or if the structural design had been based on narrow margins of safety in strength and durability of the concrete. If not more than ten or a dozen specimens are to be tested at any one age, the requirements are usually expressed in the form of the conventional absolute minimum strength, but if more than about 20 or 30 results are likely to be obtained other methods of stating the requirements may be more appropriate.

The specification of compressive strength in terms of an absolute minimum value at a certain age has the merit of simplicity and is probably adequate for many jobs. Under this system the contractor produces concrete which may seem adequate to him, while a supervisor is entitled to sample and test any material, but particularly any which appears to him to be of doubtful quality; the sampling is not necessarily random and, in fact, is probably highly selective. The adequate checking of quality may thus depend on the vigilance of the supervisor and on his ability to select the samples of relatively poor quality.

Even though the sampling for checking compliance will be so selective, the contractor designing a mix to comply with this type of specification may have, in effect, to decide what proportion of the concrete produced he can afford to have fall below the minimum, and this will depend largely on the attitude of the supervising authority's representative towards the test results. If he is very pedantic and would automatically require all concrete represented by a result falling below the minimum to be cut out, the contractor may choose to base his mix design on a chance that only 1 in 500 or 1 in 5,000 batches (not samples) of concrete will fall below standard; for random sampling $\bar{x} - 3\sigma$ (the value corresponding to about 1 in 750 results falling below it) is sometimes regarded as the effective minimum value for a set of results in practice.[17] But if the engineer is aware of the difficulties of testing and is prepared to consider the causes of each low result and only order cutting out if the evidence warrants it, the contractor may design a mix on a chance of only 1 in 100 or even 1 in 25 batches having a strength below the minimum.

This simple system of specifying an absolute minimum strength has the serious disadvantage that the greater the number of samples submitted for test the greater is the chance of finding one that will fail to comply with the requirements of the specification: even though the sampling is not random, the effect is similar to that illustrated in fig. 2. The contractor will therefore be very reluctant to have his material sampled and will tend to regard testing with grave misgivings; a system of checking the control which unnecessarily antagonises the producer is unfortunate, to say the least.

## ENFORCING THE SPECIFICATION

If the supervisor's suspicions appear to be confirmed by the compressive strength of a sample being below the specified minimum, some action has to be taken to enforce the specification requirements. As the fate of the 'consignment' represented by the sample, which could be a large section of concreting, may be at stake, it is usually worth while checking whether the low result might have been due to some fault at any stage of the testing—a long and complicated procedure which is not discussed here. If, then, the concrete is found to be sub-standard, further action has to be taken to ensure that the requirements of the specification will be satisfied. If no action is taken at all, it would seem that the supervising authority is under a moral, if not also a legal, obligation to accept the responsibility for the quality of the work and should not expect the contractor to bear the cost of any subsequent repairs which may be necessary because of the low standard of the original work, unless there is some over-riding agreement to the contrary.

The action to be taken will not be considered in detail, but some reference must be made to it because there is little point in drafting any specification if its requirements cannot be enforced. It is sometimes thought that the desired result can be obtained by adopting a 'pound of flesh' attitude to cutting out work represented by sub-standard results, but this is not necessarily so, as pointed out later. On the other hand, a system of payment by results might provide adequate incentive for the contractor to comply with the specification; there may be some difficulties in applying such a system in practice, but it is understood that work has already been carried out to a specification which stipulated a reduced rate of payment for work represented by results below a given limit.

## REPLACEMENT OF DEFECTIVE CONCRETE

The replacement of defective concrete can usually be achieved without undue difficulty (although not necessarily expense) if it occurs in precast units and paved areas, but the operation may be extremely complicated if the concrete occurs in structures, and some system of reinforcing the structure may have to be applied. However, a number of considerations ought to be taken into account before drastic action is demanded. The approach is dependent to some extent on whether the compressive strength is important for structural reasons or for ensuring satisfactory durability.

It may well be that any remedial action will have undesirable effects on characteristics of the structure other than strength or durability: for example, the replacement of sections of a road may result in a considerable worsening of the riding quality of that part of the highway, or the renewing or buttressing of a structural member or the application of a protective coat to defective concrete may seriously spoil the appearance of the structure as a whole. Any undesirable effects of taking remedial action should therefore be balanced against the improvement to be achieved, and it may often be found advisable to leave things alone.

If the compressive strength of the concrete is specified to ensure structural stability only, there are a number of sobering thoughts which point to the absurdity of demanding automatically the cutting out of concrete for the sake of the strength of the structure. In the first place, the compressive strength of concrete increases with age and a structure is often not subjected to the full design load at the age at which the strength is specified; the quantitative significance of this effect may be judged from table 9 of CP 114 (1957).[18] Secondly, concrete creeps under load with the result that excessive stresses at one section are transferred to the surrounding concrete which is not so highly stressed and the stronger concrete can help the weaker. Thirdly, for a number of reasons discussed in chapter 3, the safety of the structure often depends much less on the quality of the concrete than on many other factors.

The arguments in the preceding paragraph do not apply when the compressive strength is to be taken as an indication of the durability of the concrete, and the replacement of the sub-standard concrete may then seem to be very desirable. But, whatever the argument for dealing with a sub-standard batch in any way, there is always the disturbing thought that only a small proportion of the batches of concrete used are ever tested; furthermore, the supervisor may well have missed sampling much concrete that was equally poor. It has sometimes been said that a low result is as much a reflection on the ability of the clerk of works to supervise the job as on the general foreman to organise it, and under such circumstances he may be loath to sample batches of doubtful quality and have their strength put on permanent record. This unknown doubtful concrete will be in unidentified parts of the structure; no action can ever be taken to deal with it and no one will ever lose any sleep over it! On the other hand, the defective sample of concrete may well have been taken to represent a much larger volume of material than the batch from which the sample was taken; much of this 'consignment' might well be of better, and adequate, strength and have to be sacrificed for the sake of the defective batch.

It seems, therefore, that it is unrealistic, having fixed a minimum strength for the concrete, to enforce that requirement only by demanding that action be taken to rectify the concrete in the structure represented by any sub-standard specimens. This method may be justified on the grounds of simplicity for small but important works where there is constant expert supervision and only a few samples are taken deliberately from what appears to be the worst concrete, because it will at least avoid accepting concrete which is known to be below standard. It has been suggested[19] that, for some types of job, any concrete which looked to be of doubtful quality would automatically be rejected for use in the work, but that it would be sampled and tested and then paid for, even though it had not been used, if the material was found to comply with the requirement of the specification; the cost of testing should be compared with the value of the batch of concrete before adopting this procedure. But wherever possible more logical ideas should be incorporated into the specification to draw attention to the need for all the concrete, and not just the part tested, to be up to some acceptable standard.

## STATISTICAL CONTROL SPECIFICATION

If the specification is to provide control over the quality of all the concrete that is produced, it must relate to samples taken at random to represent the overall variation in the concrete; the test results are then examined collectively rather than individually. This technique, based on the statistical approach, may be used for specification purposes.

Unfortunately, a statistical examination of results is more effective the greater the number of results available. A statistical control specification, if it is to serve a useful purpose, is therefore likely to involve more testing than the simpler requirement of a definite minimum strength. This may be a distinct disadvantage, and for many jobs it would be more economical to use a richer mix to give a greater chance of complying with the requirements of the specification than to go to the expense of the necessary amount of testing; this, of course, would not apply if the concrete were being supplied from a ready-mixed concrete plant which could operate on a similar basis to a large contract. On the other hand, a statistical control specification can lead to considerable economy on large jobs where the cost of the concrete is an important part of the cost of the work. Whatever the class of work to be undertaken, an appreciation of the statistical type of specification should help to provide a better understanding of the interpretation of other specifications.

As it is obviously undesirable that compliance with the specification should depend on the number of samples tested, especially when compliance is made easier by reducing the amount of testing, a specified minimum strength should be associated with a stated proportion of test results, or, expressing the idea another way, with a probability that not more than one in so many results will fall below that value. If the average value ($\bar{x}$) and standard deviation ($\sigma$) of the compressive strength results available are calculated, it is possible to estimate the level of strength below which the required proportion of results, taken from the same run of production, might be expected to fall; taking the appropriate value of 'k' from fig. 3, this estimated minimum would be $\bar{x} - k\sigma$. This minimum, which for practical purposes is independent of the number of results used in its calculation, would then be compared with the specified minimum value.

The specification of a minimum strength in terms of a probability that a stated number of results will fall below that value implies also that there is the same chance of failure in all batches made, whether they have been tested or not, if the sampling has been random. There need be no concern, therefore, about batches of concrete which, if they had been tested, would have failed to reach the minimum strength, because their presence is recognised in this statistical approach.

The enforcing of such a specification is based on a different concept than expecting the contractor to carry out any remedial works to the structure at his own expense. This type of specification does not remove the right to demand the replacement of sub-standard concrete, but, on the other hand, it does not depend on this right for its enforcement. It can create a natural financial incentive for the contractor to comply with its requirements,[20] and by introducing logical cash penalties according to the statistical characteristics of the results the contractor can be given a financial inducement to adopt a high degree of control.[21]

The specification requires the keeping of a regular check on the quality of the concrete. Under favourable conditions this can lead to the forecasting of the need for a change in the concrete mix before unsatisfactory concrete has been produced. Under less favourable conditions the results still enable one to make a reasonable assessment of the seriousness of any lack of compliance with the specification upon which further logical action can be based; this assessment involves an appreciation of the statistical aspects of structural safety.

CHAPTER THREE

# Structural Safety

## MEANING OF SAFETY

We have seen in chapter 2 that a specified 'minimum strength' has very little meaning unless we associate with it an indication of the probability that some results will be lower than this limit. In the past we have traditionally designed our structures with a so-called 'factor of safety'; this factor of safety may be defined as the minimum strength of the material accepted in a member divided by the maximum stress that that member will be designed to carry, irrespective of whether they both occur in the same part of the member. However, this definition leaves a lot to be desired if the term 'minimum strength' by itself has little meaning; furthermore, we may apply the same statistical arguments to show that the term 'maximum stress' is equally ambiguous without a probability that some stresses will exceed the value.

This being so, we must think in terms of the probability that the strength of the material in any given part of the member is less than the stress in the same part. Such an occurrence may give rise to a local failure in the member, but whether failure actually occurred, either in the member or in the structure, would depend on the ability of the concrete to accommodate itself to the excessive load by redistribution of stress. The problem of safety is thus one requiring the statistical approach.

The concept of safety in statistical terms was developed considerably during the second world war in certain branches of engineering, and interest in applying this approach to structural engineering came a little later.[22-25] The trend was probably less obvious to the average practising engineer than was the corresponding development in the use of statistics for controlling the quality of the concrete produced on the job. Nevertheless, some evidence was being collected on the variability of loading to which structures were subjected, and the results of structural research were being treated by statistical methods to investigate the chance that design calculations used in practice would underestimate the stress causing failure in a member.

The European Committee for Concrete[25, 26] has played an important part in organising tests and correlating the results and in making recommendations for incorporation in Codes of Practice. The introduction into British Standard Code of Practice CP 114[18, 27] in 1957 of the load factor method of designing beams and slabs was one result of these developments. The way had been paved to some extent by the publication of a report on structural safety prepared by a committee of the Institution of Structural Engineers,[28] published in 1955, and by the Symposium on the Strength of Concrete Structures organised by the Cement and Concrete Association in 1956.[27, 29-31] A review of some of the papers of interest on this subject is given in Appendix 3.

No attempt will be made to discuss here the long philosophical and sociological arguments affecting the probabilities against failure that are required for designing different types of structures if their failure would involve loss of life and have great economic repercussions. This is a specialised subject which involves such difficult concepts as the assessment of the value of life in terms of money, the risks people are prepared to take without becoming worried and the useful life that structures ought to have. These are problems for the designer of the structure and he will be well advised to consult some of the specialised papers on the subject, to which reference is made. The use of statistics in assessing design criteria is becoming of greater importance, as is shown by some of the more recent reports, but there is a danger that wrong conclusions will be reached unless the mathematics is tempered with engineering experience and common sense.[32] The present discussion will be limited primarily to aspects of construction which will affect the ultimate safety of a structure once it has been designed whatever the arguments lying behind the criteria of safety incorporated in that design.

## PROBABILITY OF FAILURE

The idealised design problem related to the failure of unreinforced concrete in direct compression is illustrated in fig. 4. The horizontal axis in the diagram is one of strength of material or stress to which the material is subjected; both are reduced to a common scale because the maximum stress which will be withstood by a standardised test specimen is usually different from the maximum stress which would be withstood by the same concrete under the particular conditions of loading in the member. For our present discussions it is immaterial whether we

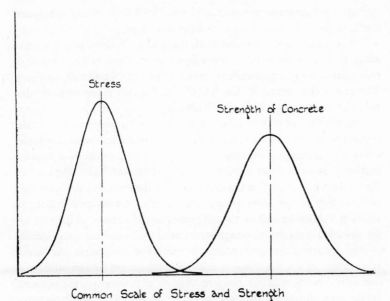

Fig. 4—Idealised design problem for probability of failure

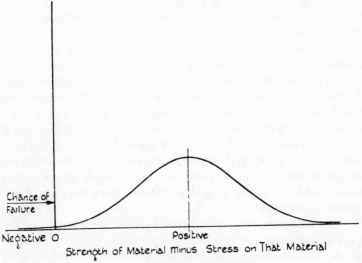

Fig. 5—Probability of failure

convert the compressive strength, as measured on standard test specimens, to the equivalent design stress, or vice versa.

The variation in compressive strength of concrete of nominally the
same quality is assumed to be subject to random variations and is
represented by the probability curve on the right-hand side of fig. 4.
The probability curve on the left of the diagram represents the distribution of stresses within the structure.

As with the quality of concrete, there is quite a wide variation in the
stresses to which various parts of the structure are subjected. Stresses
below the maximum working stress often occur, sometimes because
parts of the structure are deliberately designed to carry low stress rather
than introduce a change in the dimensions of the members or sometimes
because the design assumptions are unduly conservative. But, in
addition, there may well be a small proportion of stresses which exceeds
the so-called 'maximum' design stress used in the calculations because
of inaccuracies in the assumptions on which the calculations are based.
In practice, of course, stresses above the maximum design stress can
also occur because of overloading due to unforeseen circumstances.
While some forms of overloading, such as excessive wind or snow
loads, may well come within the range of random occurrence, other
types of overloading occur because of a deliberate increase in the
loading, which may then be systematic rather than random; such forms
of overloading would not be represented by the idealised normal distribution shown in the diagram.

Local failure may therefore be seen to be possible at any stress if
the stress in a particular part of the structure exceeds the strength of the
material in that part. We can estimate the probability of failure at each
of a range of stresses by multiplying the probability that stresses will
exceed a particular value by the probability that material would have a
strength less than the strength equivalent to that stress.[23] Alternatively,
we may visualise the single probability curve for the values obtained
when the stress in any one part of the structure is subtracted from the
strength of that part, where both stress and strength are given on a
common scale, as in fig. 5, and we then estimate the chance that
negative values will occur.[33]

It must be pointed out immediately, however, that the probabilities
of failure calculated in these ways should not be taken at their face value
because the probability of failure is usually so remote that it cannot be
estimated accurately from the amount of data normally available and

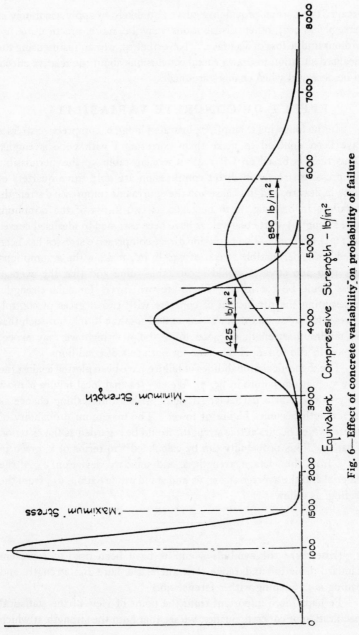

Fig. 6—Effect of concrete variability on probability of failure

D

because the normal probability curve is unlikely to apply accurately at extreme values; other distributions may be more practical to use without undue loss of accuracy.[34] Nevertheless, we can justify using the idealised situation to draw general conclusions about the relative effects to be observed when changes are made.

## EFFECT OF CONCRETE VARIABILITY

In the following example,[35] illustrated in fig. 6, compressive stresses have been doubled to make them correspond with cube strengths. (This ratio is based on CP 114[18] where, in table 5, the permissible compressive stresses in direct compression are only three-quarters of those in flexure, and in Clause 306 the equivalent compressive strength, at failure in bending, is to be taken as two-thirds of the minimum cube strength.) Particular values have been assigned to idealised curves for 1:2:4 concrete. The curve for direct compressive stresses has been based on a 'permissible stress' of 750 lb/in², making the assumptions that 1 in 100 stresses would exceed this value and that the average stress would be 500 lb/in². There are two curves for cube strengths representing the production of concrete with two degrees of control, represented by standard deviations of 850 and 425 lb/in², but such that the minimum strength of 3,000 lb/in², below which we may expect 1 result in 100 to fall, is the same for both sets of conditions.

The theoretical probabilities of failure have been plotted against the cube strength at failure in fig. 7. We can see that local failure is most likely to occur at a particular strength, with a diminishing chance as the strength becomes higher or lower. The maximum probability of failure, which occurs at this strength, would be regarded as the criterion of safety. This probability can be calculated[36] in terms of k, given in fig. 3, from the average strength, $\bar{s}$, and standard deviation, $\sigma_s$, of the material and the average stress, $\bar{w}$, and standard deviation, $\sigma_w$, from the loading, as follows:

$$k = \frac{\bar{s} - \bar{w}}{\sqrt{\sigma_s^2 + \sigma_w^2}}$$

(Elsewhere the symbols s and w have been used to represent standard deviation and range, but their use here for strength and loading is in keeping with reference 36.)

Perhaps more important from the point of view of the statistical specification of concrete quality, we see that both the strength at which failure is most likely to occur and the maximum probability of failure are appreciably different according to the degree of quality control of the material: the lower standard deviation leads to a considerably lower chance of a failure despite both levels of control giving concrete which just complies with a requirement for minimum strength.[37, 38]

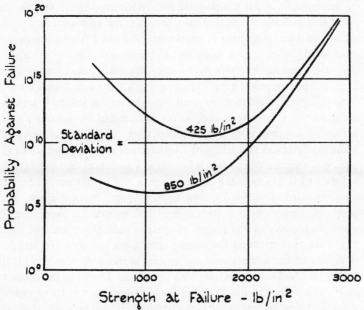

Fig. 7—Theoretical probabilities against failure

As one might expect, an increase in average strength would also decrease the chance of failure for any given degree of variation in quality. But for the normal risk of failure taken in the design of structures a change in the standard deviation of the compressive strength of random samples is more likely to have a significant effect on the probability of failure than is the corresponding change in average strength which would be needed to give the same relationship to the specified minimum strength. This may be seen from fig. 6 by assuming that unsatisfactory concrete had the larger of the two standard deviations illustrated, but had the average strength corresponding to the concrete with the lower standard deviation. If the production methods were corrected so that the concrete would just comply with the minimum strength specified previously, a change in only the standard deviation or in only the average would result in the two curves already shown, with the effect illustrated in fig. 7.

So far in this discussion we have assumed that the variation in strength of the concrete in the structure was the same as the variation in the strength of the compressive test specimens, but in practice the latter also includes the testing error which is not relevant to structural safety. If the testing error is known (it will be in a form equivalent to a standard deviation or a coefficient of variation) it can be 'subtracted' from the standard deviation for the compressive test specimens. However, it must be pointed out immediately that the term 'subtracted' is used figuratively, and that one must not subtract one standard deviation from another; the appropriate effect is obtained by subtracting the square of the one standard deviation from the square of the other standard deviation and taking the square root of the difference. (The square of a standard deviation is known as a 'variance', and we therefore subtract variances and take the square root of the difference to find the new standard deviation.)   The testing error for specimens made properly, according to a standardised test procedure, should be the same irrespective of the degree of quality control on the job; hence the effect of eliminating the testing error from the observed standard deviation of the compressive test results is more significant with the higher standard of control than with the lower. The effect on structural safety of a change in the degree of control, expressed in terms of standard deviation of test results, is therefore, if anything, even greater than has been indicated in fig. 7.

The problem of structural safety in relation to concrete strength has, of course, been very much over-simplified in considering unreinforced concrete in compression.   In reinforced or prestressed concrete the variation in the strength of the reinforcement may have a more important effect than that of the concrete, and there are other factors to be considered, some of which will be discussed later, but a similar approach can be used.

The principle embodied in fig. 4 is also relevant to the specification of strength to ensure durability because the aggressiveness of destructive agencies varies from place to place in the structure and could no doubt be represented by a probability curve. The example has been used primarily to illustrate the use of statistics in studying structural safety and to draw attention to the important effect of controlling the quality of all materials and workmanship on the ultimate behaviour of the structure.

## LOAD FACTORS

In the ultimate load design of concrete structures where conditions are favourable to the relatively accurate assessment of the statistical distribution of both the load and the strength of material, load factors varying from 1·25 to 2 may be adopted according to various considerations affecting the consequences of failure. For this purpose, the load factor may be taken as the ratio of the strength of material below which 1 in 100 results may be expected to fall to the maximum stress above which 1 in 100 stresses may be expected to occur.[39] A load factor of 2 is normally required in Codes of Practice to provide an adequate safeguard, especially as it allows sufficient margin to permit selective processes such as the rejection of obviously unsatisfactory material and the prevention of obviously excessive loading. It is reasonable, therefore, that the British Standard Code of Practice CP 115 for pre-stressed concrete[40] should allow a load factor of only 1·5 on dead loads, while 2·5 is required for live loads; there is an overall limit of 2 on the combined load.

Not all the factors of structural safety can as yet be treated statistically; for example, the effects of fatigue and vibration, instability due to aerodynamic causes and fire damage must be treated separately. The factors which may be taken into account in choosing the load factor are: (1) the seriousness of results of failure (human or economic); (2) workmanship; (3) load conditions, including consideration given to combined loads; (4) importance of the member in the structure; (5) warning of failure; (6) loss of strength, including provision for maintenance. These factors can be weighted according to the importance of their effect on the choice of the load factor. An alternative system has also been suggested.[28]

Of these, only (2) workmanship—and to some extent (6) loss of strength—are likely to be affected by what happens while the building is being erected. In the recommendations recently published by the Institution of Civil Engineers[39] only four points out of a possible 24 are allocated to workmanship and one point to loss of strength and maintenance. The adjustment for workmanship presumably excludes the level of control of production of the concrete and the variability in the quality of the steel; both are governed by statistical requirements for minimum strength. It would therefore appear that it refers to such activities as the erection of the shuttering, so that the dimensions of the members are within a close tolerance and the fixing of reinforcement so that its position is within acceptable limits.

There is an interesting commentary on the relative importance of the variation in the quality of concrete and the variation in the dimensions of the member and the position and load-carrying capacity of the reinforcement on the variation in strength of under-reinforced floor slabs.[30] Even with poorly controlled quality of concrete, represented by a coefficient of variation of 27 per cent (or a standard deviation of 1,000 lb/in² with an average strength of 3,700 lb/in²), the variation in the quality of the concrete made a negligible contribution to the total variation in the strength of the suspended floors, the remainder being divided about equally between the variation in the strength of the reinforcement and the effective depth of the reinforcement. Under such conditions it may well seem, therefore, that undue emphasis is being placed on controlling the quality of the concrete produced on the site and that some of the effort expended in this way might be better redirected to examining statistically the variation in the other critical factors.[41] Nevertheless, the situation may not be so serious because, even though it is rare for samples of the steel delivered on site to be taken and tested, engineers have the manufacturer's guarantee of quality. Also, there may be little need to examine the position of formwork and reinforcement statistically because they are open to 100 per cent inspection and can be checked selectively.

The report of the Institution of Civil Engineers[39] points out that both CP 114[18] and CP 115[40] include further safeguards in the design of over-reinforced sections, in which the quality of the concrete is more critical and the warning of failure less pronounced.

## STATISTICAL BASIS OF LOAD FACTORS

The limits corresponding to a probability of 1 per cent of the results of strength tests falling below the minimum and of 1 per cent of the stresses exceeding the working maximum seem to have been generally accepted since the early days of consideration of the statistical aspect of structural safety; but about 1 in 45, or 2–2½ per cent, has been accepted for strength results in recent Codes of Practice.[40, 42] Whether the probability of 1 per cent is appropriate to the design loading or not, it seems that the probability to be associated with the calculation of a statistical minimum strength of concrete for specification purposes needs to be re-examined carefully.

Reference has already been made to the dangers of trying to estimate the minimum value corresponding to low probabilities when only a few results are available. From this point of view it would be desirable to express the specification in terms of a minimum strength for which there is a probability of 1 in 25 or even 1 in 10 results falling below this value. On the other hand, we must consider what probabilities would be appropriate in the light of past experience embodied in documents such as Codes of Practice.

CP 114 (1957),[18] for normal structural concrete, for example, does not acknowledge that any proportion of the results might be tolerated below the minimum. However, we can obtain an implied value for the proportion of results which might be expected to fall below the minimum by examining the relationship between the specified minimum strengths at 7 or 28 days and the average strengths to be expected when using the recommended mixes; we must assume a degree of control appropriate to the workmanship clauses of the Code and values for the water/cement ratio which might be used in practice to attain suitable workabilities with the mixes specified. But when we do this the range of probabilities is very wide (roughly from 1 in 10 to 1 in 10,000), so it seems that CP 114 is of little help in assessing an appropriate proportion of results which might be expected to fall below the minimum specified. CP 115[40] states quite clearly that, for the special method of quality control, the mix should be designed so that the average strength exceeds the specified minimum value by twice the expected standard deviation; this implies that about 1 in 45 of the results (each is the average of three cubes) may be expected below the minimum.

However, a survey has been made of the cube results obtained in recent years on a large number of sites in Great Britain and an analysis of 449 sets of results gives some indication of what actually happens in practice. These cube results were not necessarily taken deliberately at random with a view to checking compliance with a statistical specification; but, because all the sets of results examined consisted of relatively large numbers of results (never less than 25, and usually between about 50 and 100), it is reasonable to suppose that all the results from one site taken together could be assumed to represent a random sample, even though minor systematic trends probably occurred during the progress of the work. The value of k (where the specified minimum strength = the average strength for the job – k × the standard deviation for the job) has been calculated for the results from all the jobs for which the specified minimum strength was known. The average values of k and the standard deviations of k for strengths specified at 7 days and 28 days and for different classes of work are summarised in table 1. It will be seen that the overall average value

### TABLE 1

**Values of k appropriate to specified minimum strength for 449 sets of cube results**
k is obtained from the formula:
Specified minimum strength = average strength for job—k × standard deviation for job

| Class of job | | Speci-fied age (days) | No. of results | Average of k | Equivalent chance of failure of individual results | Standard deviation of k | Proportion of jobs where average ≯ specified minimum | |
|---|---|---|---|---|---|---|---|---|
| | | | | | | | Esti-mated† | Actual |
| In-situ reinforced concrete | Control* A | 7 | 19 | 1·78 | 1 in 27 | 1·43 | 1 in 9½ | 1 in 19 |
| | | 28 | 28 | 2·08 | 1 in 55 | 1·06 | 1 in 40 | 0 in 28 |
| | Control* B | 7 | 68 | 1·90 | 1 in 35 | 1·18 | 1 in 19 | 3 in 68 |
| | | 28 | 109 | 1·83 | 1 in 30 | 1·09 | 1 in 21 | 3 in 109 |
| | Control* C | 7 | 39 | 1·43 | 1 in 13 | 0·97 | 1 in 14 | 3 in 39 |
| | | 28 | 39 | 1·69 | 1 in 22 | 1·13 | 1 in 15 | 2 in 39 |
| | Total | 7 | 126 | 1·74 | 1 in 24 | 1·17 | 1 in 15 | 7 in 126 |
| | | 28 | 176 | 1·84 | 1 in 30 | 1·10 | 1 in 21 | 5 in 176 |
| Paved areas .. | | 7 | 28 | 1·86 | 1 in 32 | 1·35 | 1 in 12 | 2 in 28 |
| | | 28 | 45 | 1·98 | 1 in 42 | 0·97 | 1 in 50 | 1 in 45 |
| Precast concrete .. | | 7 | 29 | 2·46 | 1 in 140 | 2·62 | 1 in 6 | 2 in 29 |
| | | 28 | 25 | 2·59 | 1 in 200 | 2·28 | 1 in 8 | 1 in 25 |
| Dam construction | | 7 | 15 | 0·96 | 1 in 6 | 0·92 | 1 in 7 | 2 in 15 |
| | | 28 | 5 | 0·49 | 1 in 3½ | 1·04 | 1 in 3½ | 1 in 5 |
| Total .. .. | | — | 449 | 1·86 | 1 in 32 | 1·38 | 1 in 11 | 21 in 449 |

*These standards of control are basically those adopted in the report 'Quality of Concrete in the Field', prepared by a Committee of the Institution of Civil Engineers[46].

†Estimated on the assumption that k is distributed 'normally': the lack of agreement with the observed proportions indicates that the distribution of k was probably skew, tailing off to higher values.

for k for all the jobs corresponds to a chance of about 1 in 30 results falling below the specified minimum strength.

The analysis was arranged to give the statistical properties of the results as though each cube had been taken to represent an individual batch even though originally they may have been taken in groups from the same batch. It would be expected that the standard deviations calculated for the results taken as individual cubes would be higher than those which would have been obtained if the results had been treated in groups of three and sampled from different batches according to the requirements of Clause 601 c of CP 114 (1957).[18] It was found that, when the results for jobs, where the individual results could be grouped naturally in threes, were treated in these two ways, the standard deviations were little different; but, occasionally, with excessive testing error and low variability of the quality of the concrete, the difference could be as much as 15 per cent.[43] The average proportion of results expected to fall below the minimum for this extreme example, treated as required by the Code of Practice, would therefore be of the order of 1 in 50, the value from CP 115,[40] but the average chance of results occurring below the specified minimum strength would be little more remote than 1 in 30.

Table 1 also shows that there was a tendency for the proportion of results below the specified minimum to be smaller for strengths at 28 days than at 7 days. For *in situ* reinforced concrete there was also a decrease in the proportion as the standard of control improved, but the trend was not very marked: the average proportion of results falling below the specified minimum from these jobs was 1 in 25 to 1 in 30. It is interesting to note that for precast concrete the average proportion of individual results below the specified minimum was very small, but that the range of proportions applicable to this type of work was relatively wide. A high proportion of the results for concrete for dams appears to have fallen below the minimum.

American Concrete Institute Committee 214[44] recommends that the proportion of results expected below the specified minimum should range from 1 in 20 to 3 in 10, according to the importance of the work. Other American experience[45] suggests that the proportion of results lower than the specified minimum cylinder strength at 28 days should be between 10 and 20 per cent. In fact, it is generally of the order of 1 out of 9 to 1 out of 4:[17] this is based on general experience of the average strength and variation in strength considered to be acceptable in relation to the specified strength and which, it seems, is implicit in the fixing of the specified strength appropriate to the design stresses of the Building Code. The author[17] recommends that the specified strength be regarded as the value below which about 16 per cent (1 in 6) of the results may be expected to occur; this minimum strength corresponds to $\bar{x} - \sigma$ and is therefore convenient to use. In the work reported by Johnson,[30] already referred to, about one result in four occurred below the specified minimum value.

Perhaps the most striking observation from the British survey, however, was that the average cube strength for a job did not exceed the specified minimum for that job in about 5 per cent of all the work investigated. The large number of structures which must have been built with concrete having an average strength below the specified minimum strength, especially on sites where there was much less supervision than on those examined in the survey, will doubtless give moderately satisfactory service.

It would seem, therefore, that, while structures are designed on criteria similar to those used in the past, there can normally be little justification in framing the compressive strength requirements for a statistical control specification in terms of a probability that as few as 1 in 100, or even 1 in 50, individual test results may be expected to fall below the minimum. Serious consideration should therefore be given to selecting the proportion of results to be expected below the specified minimum appropriate to the importance of the particular structure to be erected, and to the choice of load factor.

# CHAPTER FOUR

# Statistical Control Specification

## CHARACTERISTICS OF THE SPECIFICATION

The simplest method of specifying the quality of concrete, on a basis of tests to assess the level of control, is to state a working minimum compressive strength and its statistical significance in terms of the proportion of individual results that can be expected below the value. Statistical calculations can then be made from time to time on sets of results to check whether they indicate that the concrete quality complies with the specification requirements. If the characteristics of the results become such that the estimated minimum strength would be expected to fall below the specified minimum, the requirements of the specification must be enforced by taking action to raise the average strength, by improving the type of mix used, or to lower the standard deviation, by making very definite changes in the control technique.

Analysis of site cube results has shown that the standard deviation is affected much more by the amount of care and attention given by the site personnel to the various stages of the production of the concrete than by the type of equipment available for doing the work.[43] It is not easy, therefore, to estimate accurately the standard deviation that will be obtained, nor is it wise to assume that the failure of any concrete to comply with the requirements of the specification because of excessive variation can be overcome by small improvements to the plant unless, of course, obvious faults have been detected in it. In general, therefore, the contractor is more likely to comply with the requirements of the specification again by modifying the mix proportions, to increase the average strength, than by making a change in control procedure, aimed at reducing the standard deviation, although the effect on the actual safety of the structure may be relatively small.

The statistical approach does not eliminate the problem of uncertainty associated with testing errors, as discussed in chapter 2, and with random variability, although it is less acute for the statistical characteristics calculated from a set of results than for an individual test result. There is still a chance that a contractor may be penalised because the luck of the draw in sampling gives results which suggest that concrete which is satisfactory under ideal testing conditions is below standard, or that the engineer may have to accept concrete below the ideal standard because the test results indicate that it complies with the requirements; the ideal here is the result that would have been obtained if every batch had been tested to provide a precise estimate of the quality of production. It would be unwise, therefore, to try to design a mix too closely to the limits until some experience of this type of specification has been obtained, particularly in view of the difficulty of forecasting the standard deviation of the results for the whole of the work, even if the value is based on past experience.

## MEANING OF STATISTICAL MINIMUM

The uncertainty of the statistical estimates referred to is due entirely to taking random samples and not to any change in the level of control of the quality of the concrete: in other words, if several sets of the same number of specimens had been taken at random from the concrete made on a uniformly controlled job, different batches being sampled for each specimen, the values of the average, $\bar{x}$, and the standard deviation, $\sigma$, for these sets would have varied. In the same way, therefore, the estimate of the statistical minimum strength of the consignment obtained from $\bar{x} - k\sigma$ for a particular set of results will depend on which of the batches of concrete had been included and the actual value of $\bar{x} - k\sigma$ for any one set might be higher or lower than the value which would have been obtained if every batch of concrete had been tested.

The expression of the minimum strength as a simple probability that one result in so many will fall below the value implies, therefore, that there is a 50/50 chance that in any set of results there will be more than, or less than, the stated proportion falling below the minimum when the quality of the concrete as a whole just complies with the requirements. For example, if the strength were to be specified in terms of a minimum below which we would expect 1 in 25 results to fall, and concrete was produced just complying with this requirement, it is quite possible that, out of 250 test results obtained from random samples, there may be 9 or 11 results below the minimum, even though we would have expected there to be 10. Similarly, out of 25 random results, we may have 0 or 2 (or even 3 on rare occasions) below the minimum, although the production was still in control at the same level. Since, in practice, we are dealing with limited numbers of results per set and may be counting the number of individual results below the limit (which can only be an integer), it is necessary to draw a distinction between actual numbers and proportions: for sets of 25 results, as in the example above, the most frequent number of results below the limit will be 1, with 0 and 2 results occurring less frequently.

Some authorities have attempted to make the traditional specification requirement independent of the number of samples tested by specifying a minimum compressive strength below which not more than a stated proportion of results will be permitted to fall: *e.g.* they require that not more than one actual result in 25 shall fall below the stated value. However, this is still illogical statistically. With material that just complied with the intention of a specification, in which the limit would be set by a minimum below which we would expect 1 result in 25 to fall, there would still be a chance that an extra odd low result, falling below the '1 in 25 minimum', would lead to it being unjustifiably condemned; similarly, materials which ought to be better might escape because the lowest results were not quite low enough for more than one to fall below the minimum. In fact, the whole acceptance or rejection of a series of concreting operations could depend on the lowest one or two results out of the 25 and all the other results would have had no significance.

The minimum, which is to be related to the specified minimum, should be a calculated value—and not an individual result—as already suggested. Provided all the results are used in the calculation, there is no ambiguity in the calculation of the minimum value; the uncertainty lies only in the result that might have been obtained if some other samples had been tested.

## SAMPLING AND TESTING PROCEDURE

The uncertainty associated with a statistical control specification can be reduced by having an adequate number of results in the set to be examined statistically. On the other hand, there is a need to limit the number of results in the set to avoid too much delay in discovering whether remedial action has to be taken. The specification should therefore require that enough specimens are made and tested, but in such a way that results are obtained as quickly as possible to enable the level of control to be established with reasonable accuracy in a short time. This can be encouraged in three ways:

Specimens should be made more frequently in the early stages of a job to establish the general level of control, and then less often, but at a rate sufficient to ensure that the desired level is being maintained; sometimes preliminary concreting operations can be used as a full-scale trial for more important work to follow without undue extra expense.

The minimum strength can well be specified for tests at 7 days, or even earlier, rather than the more conventional 28 days,[46] and the specified minimum value related to the level of strength required for safety or durability of the structure.

Specimens should be treated individually so that each one, for testing at a particular age, represents a separate batch of concrete.

Reference was made in chapter 2 to the attempts which have been made to obtain a warning that unsatisfactory material was being produced earlier than would be possible with normal curing methods. It may well be that the variation in the results obtained from the accelerated test could be used as a measure of the variation in the results that would have been obtained from the same concrete in specimens cured normally. If this were so, we would expect that the average value for the accelerated test would correlate with the average value for the specimens cured normally and, similarly, that the standard deviation of the results in the accelerated test would correlate with the variation of the results obtained normally. Hence, the usual procedure for designing and controlling mixes on the basis of the average strength required to ensure a satisfactory proportion of results exceeding the minimum value could be adapted for the accelerated method of testing; this procedure would be especially useful if all the specimens for each set were sampled within a day or so of each other, corresponding to the time taken to complete the accelerated testing.

Taking the test specimens in sets of three from a batch, as has been traditional, and using the average value reduces the testing error considerably, but usually the effect of all the testing error on the observed standard deviation of the results is quite small. This may be illustrated as follows.

The value of the testing error tends to increase with an increase in the average strength and therefore is perhaps expressed more universally as a coefficient of variation rather than as a standard deviation: the testing error for individual works test cubes, made according to B.S. 1881,[2] has been found to be about 5 per cent, although higher values are obtained with leaner mixes and with mixes with aggregate of larger maximum size.[43] If, for example, the average compressive strength of concrete is found to be 4,000 lb/in² and the testing error, expressed as a coefficient of variation, is 5 per cent, the testing error expressed as a standard deviation would be 200 lb/in². The effect of the testing error on the observed standard deviation can be found by subtracting variances as discussed in chapter 3. For example, if concrete had a standard deviation, estimated from a large number of individual test results, of 600 lb/in², the standard deviation of the variation of the concrete itself, without the testing error, would then be 565 lb/in². The testing error for averages of three specimens would be $200 \div \sqrt{3}$ = 115 lb/in² and the standard deviation of the average of sets of three specimens from each batch would be 578 lb/in.²

On the other hand, there is a marked advantage in having information on three times the number of batches, by taking single specimens for each batch, without increasing the number of test specimens.[44] For example, 30 specimens may be taken individually to represent 30 batches or in sets of three from each of 10 batches. The variation in the estimate of the standard deviation based on 10 values for the average of the three results in a set taken at random from a consignment, where the observed standard deviation is 578 lb/in², is such that 95 per cent of the results would be expected to lie between about 395 and 1,055 lb/in². The standard deviation estimated from the results of 30 specimens, sampled to represent the average quality of separate batches and having an observed standard deviation of 600 lb/in², would lie, with the same probability, between about 475 and 810 lb/in².

Occasionally there may be some point in requiring that pairs of specimens (rather than sets of three) be taken from the same batch during the early stages of the contract so that the testing error can be estimated for that particular job.[20, 46] Indeed, it may be worth while for a contractor to do so on his own account to ensure that the variation in the observed test results is not being boosted unnecessarily by a large testing error, so making the concrete more liable to be unacceptable. A method of calculating the testing error from the results of pairs of specimens from the same batches is given in Appendix 1.

The specification, therefore, should normally require that the compressive strength tests are to be made on concrete sampled, at random, to represent the average of individual batches, and that one specimen from each sample is to be taken for a particular test, at an age preferably not greater than seven days. The rate of making specimens should be indicated in some measure, such as a minimum number per pour, per day or per stated volume of concrete produced, as appropriate, and should be greater in the early stages of the job and may then be reduced after about 40 samples have been taken. The specimens should be required to be made, cured and tested according to a recognised standard procedure, such as B.S. 1881,[2] to minimise testing error. In particular, care should be given to ensuring full compaction of the concrete and storage of the cubes at a constant temperature appropriate to the method of arriving at the specified minimum strength; close control of the curing temperature is especially important if the specimens are to be tested at early ages.

## NUMBER OF RESULTS PER SET

Some further appreciation of the principles involved in statistical calculations is required before one can prepare a specification of this type in detail or check whether its requirements are being met. It is necessary to examine the probable variation in the values of the average, the standard deviation, and $\bar{x} - k\sigma$, when estimated from relatively small numbers of results. It is obvious, for example, that if the results were treated in very small sets of, say, three or four at a time, the calculations of $\bar{x} - k\sigma$ might well suggest sometimes that the level of control was not acceptable by quite wide margins, and this simply because the estimates of both the average and the standard deviation for such small sets were unreliable rather than because the production of the concrete did not comply with the intention of the specification. Table 2 gives the range of values of the average, the standard deviation and $\bar{x} - k\sigma$ (for two values of k) based on different numbers of results in a set, within which the values calculated from 19 sets out of 20 might be expected when the true standard deviation is 600 lb/in$^2$, which represents a fairly high degree of control.

It will be seen from table 2 that, for concrete being made consistently with a standard deviation of 600 lb/in$^2$, there is a 1 in 20 chance that the average of a set of 10 results will be 375 lb/in$^2$ or more either above or below the true average and the same chance that the standard deviation of a set of 10 results will be below 415 lb/in$^2$ or above 1,100 lb/in$^2$. There will, of course, be only a 1 in 40 chance of the average of a set of 10 results being below 375 lb/in² less than the true average or of the standard deviation being more than 1,100 lb/in². Furthermore, the chance of any one particular set of results having both an average strength more than 375 lb/in² below the average and at the same time a standard deviation of more than 1,100 lb/in² is very much more remote than 1 in 40, as long as the results are varying consistently with a true standard deviation of 600 lb/in².

## TABLE 2

### Ranges of values of average, x̄, standard deviation, σ, and minimum value, x̄—kσ, calculated from n results

(within which 19 out of 20 values are expected to occur, when the standard deviation of the original results is 600 lb/in²)

Values of k of 2·33 and 1·28 represent respectively the probability that 1 result in 100 and 1 in 10 will fall below the minimum value

| n | x̄ | σ | | x̄ — kσ | | | |
|---|---|---|---|---|---|---|---|
| | | | | k = 2·33 | | k = 1·28 | |
| 10 | ±375 | —185 | +500 | —1,450 | +500 | —910 | +390 |
| 30 | ±215 | —120 | +210 | — 575 | +325 | —380 | +245 |
| 100 | ±120 | — 80 | +100 | — 270 | +200 | —180 | +145 |
| 1,000 | ± 40 | — 25 | + 25 | — 75 | + 70 | — 50 | + 50 |

It will be apparent from table 2 that the distribution of the estimate of the standard deviation about its mean value is skew, with the values tailing off more to higher values than to lower. As we are more interested in the maximum likely value, when calculating the statistical minimum, than in the minimum likely value, this skewness aggravates rather than helps our difficulties in overcoming the uncertainty. As the number of results in the sample becomes small, the maximum likely standard deviation of all the results from which the sample is drawn is further from the average standard deviation than is the lowest likely value of the average (calculated from the same size of sample) from the grand average. The standard deviation therefore has a greater effect on the variation of the value of x̄ – kσ than of the average, particularly if the value of k is increased to represent limits below which a lower proportion of results is likely to fall.

This effect is only to be expected because it is unreasonable to suppose that one could obtain a reliable estimate of the strength below which not more than, say, 1 in 100 results would be expected to fall

E

from the results of tests on only 10 specimens. It would, therefore, be pointless to use high values of k in an attempt to relate the values of $\bar{x} - k\sigma$ for the numbers of specimens normally tested on any job to the order of probability of interest in dealing with structural safety, which may be something like 1 in 1,000,000; for most jobs it is difficult enough to relate closely the calculated minimum to probabilities of only about 1 in 100 or less.

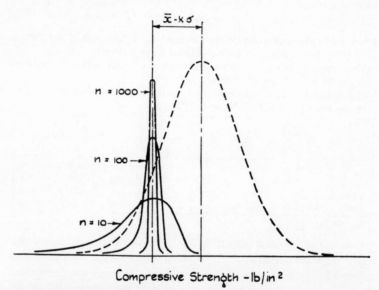

Fig. 8—**Distribution of values of $\bar{x} - k\sigma$ calculated from different numbers of results (not to scale)**

Table 2 gives values for the range of the statistical minimum value, calculated from $\bar{x} - k\sigma$ for values of $k = 2 \cdot 33$ (representing 1 in 100 results likely to fall below the minimum) and $1 \cdot 28$ (representing 1 in 10 results likely to fall below the minimum). These values have been calculated from the data given in B.S. 600R: 1947[47] for variation of the 'coefficient of displacement'; a simpler approximate formula for the calculation has been suggested,[48] but it does not indicate the skewness of the distribution as shown diagrammatically in fig. 8. It will be seen that the possible discrepancy between values of the statistical minimum as calculated from small numbers of results and as calculated from a very large number of results is even greater than for the corresponding differences with the standard deviation; as might be expected, the characteristics are worse when the value of k is 2·33 than 1·28. The greater variation of the calculated minimum than of either the average or the standard deviation is due to the fact that the value is calculated from two variables; hence the scatter in the results would be appreciably greater than for either of the two variables from which it is calculated. The skewness of the variation in the calculated minimum value is confirmed, and the tendency for tailing off to low values makes for considerable uncertainty in our estimate of the statistical minimum for the whole consignment based on the relatively small numbers of results that are likely to be available on a job.

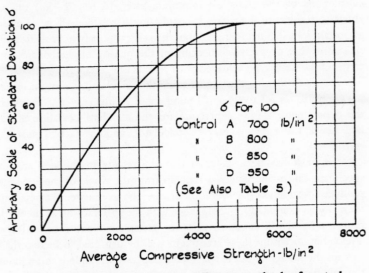

Fig. 9—Standard deviations for different standards of control

It would appear that the difficulty which we are finding arises primarily from the uncertainty in the estimation of the standard deviation from small numbers of results and we would do better if we could find a more precise estimate from some other source. The British survey[43] of over 1,000 sets of cube results, for varying types of quality control technique, has provided evidence on which the standard deviation for any particular type of control can be estimated (see fig. 9), and it appears that only about 5 per cent of the results would have a value higher than 1·45 × the average standard deviation for the particular control technique. To obtain the standard deviation for a particular job with the same degree of accuracy the value would have to be calculated from about 15 results; if the range of the estimate of the standard deviation for the concreting operation had to be only half this value, about 40 results would have to be used in the calculation.

It would seem reasonable to suppose that a contractor who had become accustomed to statistical control of the quality of concrete would be able to estimate the standard deviation likely to be obtained on a new job with considerably higher precision than the estimate based on the British survey, which includes results from all classes of work. It would seem, therefore, that there is little point in attempting to make calculations of the statistical minimum strength (*i.e.* $\bar{x} - k\sigma$) unless about 40 results are available. It is only on the relatively large or more important jobs that many more cubes than this are likely to be made and, therefore, that progressive checking of the quality of the concrete in terms of $\bar{x} - k\sigma$ can be achieved with reasonable certainty.

Before dismissing the idea of using the calculated minimum value, $\bar{x} - k\sigma$, based on relatively small numbers of results (*e.g.* 11, for which the denominator in the expression for the standard deviation is 10, and hence the value is easy to calculate), another factor must be considered.

So far it has been assumed that the results in the sets are for specimens taken completely at random from the whole of the concrete produced throughout the duration of the job; in practice the sets will normally be examined for compliance with the specification as the work progresses and will contain results for specimens made consecutively. The variation within these sets might well be expected to be less than that for the same number of specimens dispersed at random throughout the job because the consecutive sets would not reflect some of the long-term variations.

The presence of these long-term variations could be taken to indicate that the production of the concrete was not 'under control' (in the statistical sense) and that unwanted systematic variations were occurring which should be eliminated by modification of the control procedure or mix proportions. However, some of the causes of variation in the quality of concrete which are not easily controlled operate on a long-term basis; for example, changes in the characteristics of the aggregates due to wear and tear of the plant and seasonal changes in weather conditions would not be likely to be reflected in sets of 10 or 11 results. These variations, although systematic rather than random within the range of our interest, should be regarded as part of the normal hazards to be allowed for in setting the limits of the specification: hence the average variation within sets should be expected to be a little less than the overall variation for the job.

It has been found that, for 14 jobs on which about 100 to 400 cubes have been taken, the value of $\sigma$ calculated from sets of 11 cubes made consecutively was, on the average, about 15 per cent lower than

## TABLE 3

**Relationship between values of standard deviation and minimum strength as calculated from sets of 11 consecutive results and calculated from all the results for a job**

| Standard of control (see Table 5) | Overall average strength (lb/in²) | Age at test (days) | No. of sets of 11 results | Average S.D. of sets as percentage of overall S.D. | No. of times S.D. of sets exceeded overall S.D. | No. of times $\bar{x}-1\cdot28\sigma$ for sets fell below overall $\bar{x}-1\cdot28\sigma$ | No. of times $\bar{x}-2\cdot33\sigma$ for sets fell below overall $\bar{x}-2\cdot33\sigma$ |
|---|---|---|---|---|---|---|---|
| A | 1,860 | 7 | 17 | 87 | 3 | 5 | 4 |
|   | 4,890 | 28 | 9 | 82 | 1 | 4 | 3 |
|   | 5,450 | 7 | 20 | 80 | 6 | 5 | 3 |
|   | 6,630 | 28 | 10 | 86 | 4 | 3 | 2 |
| Average for A |  |  |  | 84 | 1 in 4 | 1 in 3·3 | 1 in 4·7 |
| B | 2,360 | 7 | 9 | 81 | 2 | 4 | 1 |
|   | 3,600 | 28 | 10 | 92 | 4 | 4 | 5 |
|   | 3,820 | 28 | 16 | 87 | 6 | 5 | 6 |
|   | 4,270 | 28 | 10 | 97 | 4 | 6 | 4 |
|   | 4,830 | 7 | 28 | 76 | 5 | 10 | 5 |
|   | 6,470 | 28 | 37 | 87 | 13 | 15 | 13 |
| Average for B |  |  |  | 87 | 1 in 3·2 | 1 in 2·5 | 1 in 3·2 |
| C | 2,670 | 7 | 16 | 84 | 3 | 6 | 5 |
|   | 4,140 | 7 | 14 | 95 | 7 | 6 | 5 |
|   | 7,000 | 28 | 18 | 77 | 3 | 9 | 8 |
|   | 7,880 | 28 | 17 | 76 | 4 | 5 | 5 |
| Average for C |  |  |  | 83 | 1 in 3·8 | 1 in 2·5 | 1 in 2·8 |
| Overall Average |  |  |  | 85 | 1 in 3·6 | 1 in 2·7 | 1 in 3·3 |

the standard deviation calculated from all the results for the job; in the same way the standard deviation calculated from sets of 11 results exceeded the value for the whole job in from 1 set in 2 to 1 set in 9, with the overall average of about 1 set in $3\frac{1}{2}$; the results are given in table 3. Similarly, the value of $\bar{x} - k\sigma$ calculated from the sets fell below the corresponding value calculated from all the results in about 1 set in $3\frac{1}{4}$ when k was taken as $2\cdot33$, representing a probability of 1 in 100, and in about 1 set in $2\frac{3}{4}$ when k was $1\cdot28$, representing a probability of 1 in 10. This means that, for jobs where large numbers of specimens are to be made, 1 in about 3 of the values of $\bar{x} - k\sigma$ for sets of 11 consecutive results will fall below the specified minimum if k is the factor appropriate to the proportion of the results for the whole job permitted to fall below that value. The proportion will approach 1 out of 2 (*i.e.* 50/50) as the number of specimens examined decreases and as k is reduced. This characteristic of the results does not introduce a much greater probability of values of $\bar{x} - k\sigma$ exceeding the specified value, and hence it would be unwise to base specification requirements on values calculated on less than about 40 results, as previously suggested.

## OPERATION OF STATISTICAL SPECIFICATION

It seems that for most jobs, therefore, there would be little point in specifying the quality of concrete in terms of a limiting value related to $\bar{x} - k\sigma$ calculated from relatively small numbers of results: the work involved in calculating the standard deviation would not be justified because a more precise value could be obtained from other sources. If, then, a value of $\sigma$ is to be assumed, independently of the results of tests, and the value of k would be stated in the specification on the basis of the structural design requirements, the specification can only require checking that the average strength of the concrete is adequate for the variation envisaged. This system of checking compliance would continue until an adequate number of results had been obtained from which the standard deviation could be calculated more accurately if, indeed, that many test results become available. It might, therefore, be regarded as the normal method, and any refinements incorporated at the later stage could be regarded as a supplementary method for larger or more important works.

For the purpose of the discussion we will regard a normal job as being one for which not more than 40 test results are likely to be available. In practice, however, this value of 40 could be decreased to, say, 30 or 25 if the best available estimate of the standard deviation prior to starting the job were very doubtful or could be increased to 50 or 60 if the contractor could supply convincing evidence, based on past experience of similar jobs, of the standard deviation likely to be achieved.

The normal method of specifying concrete quality would operate as follows: the specification would state a minimum compressive strength with statistical significance; the inclusion of a lower 'absolute' minimum for the job as an additional safeguard against some poor concrete being accepted may be thought advisable, but, logically, it is not necessary if the sampling is random. It would require a particular standard of quality control and either state an appropriate standard deviation or permit a value to be agreed on the basis of past experience furnished by the contractor. It would require that a mix be designed to meet the strength requirement on the basis of an average strength calculated from this value of the standard deviation for the whole of the job multiplied by the factor appropriate to the probability of results falling below the minimum. For example, the limit of quality might be set by a minimum strength of 2,500 lb/in² at 7 days, below which 1 result in 25 would be expected to fall; if the control were to be specified as control C, and the standard deviation were stated to be 850 lb/in², k would be 1·75 and the average strength 4,000 lb/in². The specification and an example are given in Appendix 2.

The test results have now to be examined to see whether the variation in compressive strength obtained indicates that the quality of concrete is likely to be such that the average strength will be 4,000 lb/in² or higher. As we have already seen, the variation in the average of a set of numbers depends on the number of values in the set and on the probability of the result lying within the particular limits. If we are expecting to use the normal method of specifying the quality of concrete for up to 40 results, it would be convenient to sub-divide the results as they occur into sets of four, for which we determine the average, thus giving 10 average values. The limits for the averages of sets of n results, below which or above which we might expect one value in 10 to occur, differ from the design average by $1·28\sigma \div \sqrt{n}$ where $\sigma$ is the value used in the design of the mix. For averages of four results the lower limit would be $0·64\sigma$ below the design average. We should therefore expect that, if the concrete were just of the required standard, there would be one of the 10 average values falling below this limit. The specification might require that remedial action be taken if more than one fell below this limit. Alternatively, we can set another limit, with a more stringent requirement for accuracy, appropriate to a probability of only 1 in 40 results being expected below this limit; this would be $0 \cdot 98\sigma$ below the design average (for the averages of four results), and remedial action might be required if any value of the average fell below this limit.

Remedial action may consist of checking whether the odd result happened to be a very low random result by checking the average value of the $n_1$ results obtained up to that stage and requiring a modification to the mix if it is less than $1 \cdot 96\sigma \div \sqrt{n_1}$ below the design average. Even if the average of the $n_1$ results did not fall below this limit, it might be as well to check whether there was any significant trend in the individual results or in the averages of four results.

The supplementary method of specifying the quality of concrete would be brought into operation at the end of the normal method. At this stage the value of $\sigma$ for the concrete already made and tested would be calculated, and new working limits for the average value would be determined on the basis of this standard deviation. If the new value of the standard deviation were appreciably different from the old, it may be appropriate to alter the design average strength, from which the limits are calculated, as well as the difference between the limits and the design average value. The checking of the averages of sets of four consecutive results would continue as before, but, in addition, it may be appropriate to require that a check be kept on the variation in the standard deviation; the method employed would depend on the number of results likely to be obtained during this supplementary stage.

If the number of test results in the supplementary stage were unlikely to exceed another set of 40, the relatively complicated calculation of the standard deviation could be eliminated by checking the variation in terms of the ranges of the sets of four results: the average value of the range of four results would be expected to be $2 \cdot 06\sigma$ and one value in about 40 would be expected to lie outside the limit of $4\sigma$. The specification might require that if any value of the range should fall outside this limit, $\bar{x}$ and $\sigma$ would be calculated for all the results obtained up to this time; remedial action would be required if the value of $\bar{x} - k\sigma$ was appreciably less than the specified minimum value. Again, it would be advisable to observe any trend in the magnitude of the ranges for the sets of four results. The control of the quality of concrete in terms of range and standard deviation is discussed further in chapter 5.

## FORM OF SPECIFICATION

The specifying authority will have to fix the minimum strength and its statistical significance on the basis of structural safety as outlined in chapter 3. It has been assumed up to now that the distribution of the compressive strength test results will be normal. But the factors used for checking compliance with the intended requirements may be affected if the distribution tended to be skew, as appears to occur with very low or very high average strengths.[43] The value of k selected for structural requirements may be modified to allow for this tendency before being written into the specification by multiplying it by a factor interpolated from table 4 if the number of results likely to be obtained on the job is sufficient to warrant such a refinement. Generally, however, the effect of skewness of distribution can be ignored.

### TABLE 4

**Adjustment of k for low and high average strengths**

| Average compressive strength (lb/in²) .. | 2,000 | 5,000 | 8,000 |
|---|---|---|---|
| Multiply k by this factor ..     ..     .. | 0·92 | 1·00 | 1·07 |

The statistical specification operates on the assumption that there will be a 50/50 chance of the calculated minimum value falling below the specified limit. This being so, the value of $\sigma$ to be assumed in the calculations for the normal method of specifying quality should be the average value likely to be obtained as assessed from the evidence available. However, some engineers may feel that this is running an unnecessary risk and would prefer to choose a higher value of the standard deviation on the understanding that the value could be modified if the supplementary method of specifying came into operation. Whatever the basis for choosing the standard deviation, the specification must make it clear that the value is to be agreed between both the specifying authority's representative and the contractor, and state the basis of this agreement to enable the contractor to tender appropriately. For example, the document should say whether the standard deviation is to be taken as the average for the appropriate degree of control, as indicated in the British survey, or as a factor times this average, and whether evidence from the contractor for similar jobs will be admitted as an acceptable basis for agreement, either at the tendering stage or after the letting of the contract.

The contractor should be given every incentive to use a reasonably high degree of control and should not be penalised, but rather should be rewarded, if, by his skill and experience, he can achieve better control than expected. For that reason the control levels for the average of the results in each group should be adjusted as soon as there is sufficient evidence to warrant doing so.

There is a danger that, with a specification involving the variation in sets of a limited number of results, as in the supplementary method, the concrete may be suspected of being unsatisfactory because of the occurrence in the set of some odd high results which increase the range or the standard deviation beyond the acceptable limit. However, no matter how disconcerting it may appear to the contractor to have to increase the average strength of the concrete because of strength results which are too high, it is nevertheless appropriate because these results, as much as low results, indicate lack of uniformity.

# CHAPTER FIVE

# Control of Concrete Quality

## SCOPE OF STATISTICAL CONTROL

Statistics can be a useful means of helping to produce concrete of a more uniform quality. It can be used either in retrospect, to indicate how various changes in methods of controlling the quality of concrete have affected the overall variation, or during the course of concreting operations, to draw attention to undesirable trends in quality or to forecast the likelihood of the quality of the concrete satisfying the requirements of a specification.

This type of analysis can be undertaken whatever the type of specification relevant to the work, provided appropriate measurements of the property specified are available for examination. However, the usefulness of the statistical analysis will depend upon the way the results can be applied and this, in turn, may well depend upon the limitations imposed by the specification. If the contractor is left with no freedom of choice of mix proportions, for example, there will be little incentive to attempt this type of control in order to make more economical concrete; on the other hand, maximum benefit will be obtained when some property, such as the compressive strength of test specimens at a given age, is specified as a statistical minimum value and when an adequate number of test results is available from samples taken at random.

The idea of controlling the compressive strength of concrete will be followed throughout this chapter, although the arguments used could well be applied to any other property measured at random. However, there is usually little point in doing so unless there is some requirement in the specification for a minimum or maximum value of that property and the effort which goes to the testing and analysis could be more than repaid by the resulting savings in the overall concreting operation. This type of control is likely, therefore, to be of more value on contracts of considerable size or importance.

Where a system of statistical control of concrete quality is in force it is important that the samples for testing should be taken at random. Simple statistical calculations are based on the assumption that the variations occur with as much likelihood of the value being above the average as below it; this assumption is found to be true enough for practical purposes in the case of the compressive strength of most concrete mixes if random sampling is attempted. This means that the samples we obtain should be as representative of the good concrete as of the bad and there should be no bias towards one particular kind of concrete. Similarly, one should not include in the calculations the results of any tests obtained from samples which have been taken at times when the concreting operation is known to be out of control and appropriate action is being taken; a particular example is the first one or two batches delivered from a clean mixer.

The attitude required for random sampling is very different from that normally associated with sampling concrete for checking its compliance with a simple specification demanding a certain minimum strength. In working to the latter type of specification there is obviously no point in taking a sample from concrete which is expected to have high compressive strength: the person responsible for enforcing the specification would be looking only for concrete which, in his estimation, was near the borderline and required a sample to be taken and checked by the standardised testing procedure to see whether his suspicions were justified. The element of surprise is then essential if the contractor is suspected of not doing the job as he ought to, and he certainly should not be given any warning which would allow him time to improve matters before the sample is taken. However, the element of surprise is not so important when samples are taken for control testing in association with statistical methods because the quality of the concrete is not assessed on any one individual result, but on all the results taken collectively. If the contractor soon learns that samples of his concrete, sufficient for one test specimen only, are being taken at frequent, but irregular, intervals he will have little incentive to do other than maintain a quality that he knows to be necessary to comply with the requirements of the specification.

If there is any doubt about the ability of the inspectors to sample at random, some 'system' should be followed. For example, for any particular pour the number of batches or the time in, say, five-minute intervals needed to complete the concreting operations should be used as a measure of progress; the batch numbers to be tested or the times at which samples are to be taken can then be selected by picking numbered pieces of paper out of a hat or using tables of random numbers published in textbooks on statistics. If this is thought to be too much bother, the numbers of samples taken each day and the times of the day for sampling should be varied as much as possible.

## EXAMINATION IN RETROSPECT

A number of compressive strength test results available at the end of a contract may contain information which, if extracted by statistical methods, would be of considerable value to the contractor for the work he undertakes subsequently. As a first step in examining these results, it is usual to construct a histogram, as in fig. 1, to give a visual indication of the variation in the compressive strength. This is now becoming common practice, and many papers on important projects described in the journals of the professional institutions include details of the concreting procedures used and histograms of the variation in strength.

Some guidance on plotting histograms may help in obtaining more effective representation of the information. In order to establish the shape of the histogram with fair precision it is recommended that the range of the strength results should be divided into about 15 intervals. The number of the intervals may be slightly greater if there is a very large number of results, but many more are unnecessary and undesirable because the larger the number of results in each interval the more likely is the height of each block to approach the true envelope. Many fewer intervals would lead to the ends of the curve being less well defined.

Having got an approximate idea of the range of the interval from this rule of thumb, the actual range should be chosen in round figures which can be related conveniently to the accuracy of reporting the results. For example, with concrete strengths recorded to the nearest 50 lb/in², a convenient interval might be either 200 or 250 lb/in². Sometimes the characteristics of the results are such that the appearance of the finished histogram will change noticeably according to the actual intervals chosen. For example, if the interval were taken to be 200 lb/in², the results might appear differently according to whether the range was taken from 4,000 to 4,200 lb/in², etc., or from 4,100 to 4,300 lb/in², etc. As a first attempt it is usually preferable to arrange for the calculated average value for the set to occur near the middle of an interval;

hence, if the average strength were known to be about 4,100 lb/in$^2$, the former set of ranges would be preferred to the latter.

In practice one would not use ranges such as 4,000 to 4,200, 4,200 to 4,400 lb/in$^2$, etc., because there would be some doubt about the appropriate interval for a value recorded as 4,200 lb/in$^2$. Instead the ranges would be taken as, for example, 3,975 to 4,175, 4,175 to 4,375 lb/in$^2$, etc., and the number of results falling in the intervals would be plotted at the average value for the possible strengths included in the interval, $i.e.$ 4,075 lb/in$^2$, which is the average of 4,000, 4,050, 4,100 and 4,150 lb/in$^2$.

If the variations in compressive strength, as recorded in the histogram, were due only to random variations typical of normal concrete quality control, we would expect to find that the envelope to the histogram would correspond approximately with the normal probability curve. However, some divergence from the smooth curve is only to be expected, particularly if the number of results available is relatively small. With a little experience it should be possible to observe immediately whether the histogram shows any unusual characteristics, such as any tendency towards there being more than one well-defined peak, any tendency for the results to be 'chopped off', $i.e.$ for there to be apparently a limit above which, or below which, results do not seem to occur, although one would normally have expected them to have done so, or any skewness of distribution.

The presence of this type of characteristic may lead us to enquire further into the reasons for its occurrence. It may be that some change has occurred during the course of the concreting operations which has made a noticeable change in the average strength which could account for there being more than one peak to the curve. It may be that selective sampling has resulted in the concrete of extreme qualities being excluded from the tests. Or it may be that skewness has been introduced because of a temporary relaxation in the supervision of the concreting operation. Slight skewness, however, is typical of compressive strengths with a high or a low average value, the results tailing off further to the low values or high values, respectively (see also chapter 4, page 69).

Another method of representing the complete set of test results is to plot the proportion of results less than a particular value against that value as shown in fig. 10; it will be seen that the line tends to become asymptotic to the 0 per cent and 100 per cent abcissae and has a double curvature. Special graph paper, known as 'probability paper', has been

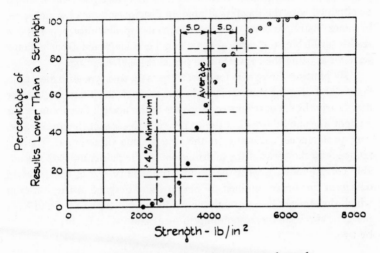

Fig. 10—Distribution of results as normal graph

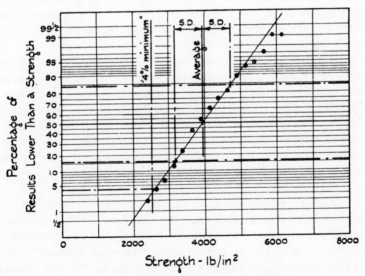

Fig. 11—Distribution of results on probability paper

prepared with a non-linear vertical scale so that the curve, which would be obtained with the theoretical random distribution on a normal scale, becomes a straight line. If properly printed probability paper is not readily available, the vertical scale can be constructed from a linear scale for k, using the conversion to percentages given in fig. 3.

By plotting the results for a set of tests on this type of paper, as in fig. 11, and drawing the best straight line through the points, we can see how nearly the set of results corresponds to the normal distribution: the straight line should intersect the 50 per cent abscissa at the mean value. Any pronounced curvature in the line in one direction indicates a tendency to skewness of the distribution. If the points indicate reasonable correspondence with the normal probability curve, the standard deviation can be determined by noting the difference in the values at which the straight line through the points intersects the 15·9 and 84·1 per cent abscissae (corresponding to $\pm \sigma$) and dividing this difference by two.

## PROGRESSIVE EXAMINATION OF RESULTS

The histogram and probability paper suffer from the disadvantage that they are not suitable for distinguishing trends which occur during the progress of the work; for this purpose it is usual to use control charts.[47, 49, 50] In a simple type of chart (which, technically, may hardly be classed as a control chart) the value of the property is plotted vertically, and the horizontal scale is used to denote progress, usually in terms of the consecutive number of the result. Points are plotted as the results become available, much as we would do with a temperature chart, but the points are not joined; thus there is only one point to one ordinate. In considering the use of this type of chart it may be helpful to think of it in relation to the corresponding histogram on its side: the histogram would be obtained if all the points on the chart were to be grouped into intervals and closed up towards the right-hand axis, as shown in fig. 12.

One way to use the chart, illustrated in fig. 12, is as follows. Suppose a concrete mix has been designed to have a certain minimum strength, in terms of a probability that one result in 100 would fall below that value, and the average strength has been estimated on the basis of a presupposed standard deviation of the results. The control chart is set out with strength plotted vertically; horizontal lines can then be drawn at the specified minimum strength, the design average

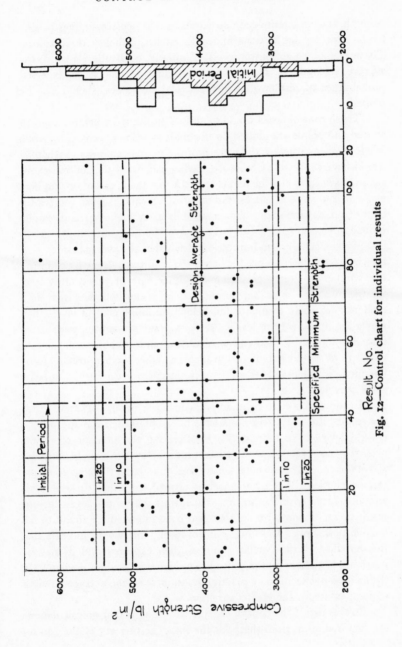

Fig. 12—Control chart for individual results

strength and at intermediate strengths, $\bar{x} - k\sigma$, corresponding to different expected proportions of results occurring below these values. These are known as control levels. Similarly, control levels above the average strength at $\bar{x} + k\sigma$, corresponding to the same expected probabilities of results occurring above these values, should also be plotted.

These control levels would be drawn on the chart before the work started and points are plotted on the chart as results become available. We then check whether the pattern of the points conforms to what we would have expected. Obviously, having calculated the control levels, we do not need to plot the points to check that the appropriate numbers occur in the various ranges, but most of us can visualise the scatter better from a graph than from a mass of data, and definite trends can be more easily observed.

We can soon get an impression of whether the average strength of the concrete being produced is below the required average by counting the number of points occurring on either side of the control level for the average. This can be checked by seeing whether there is a tendency for more results to fall below the lower control levels than above the upper control levels. If the average strength appears to be low, this should be checked by calculation and, if confirmed, the mix proportions should be modified to enable a higher average strength to be obtained; if the calculated average is satisfactory the results indicate some skewness.

In addition, we can get an impression of the amount of variability of the results after, say, 10 results have been obtained if the appropriate control levels have been drawn by checking the number of results occurring not only below the control level below the average, but also above the control level above the average. Similarly, the next 10 points can be checked both by themselves against the same control levels and also, taken with the previous 10, against the control levels for one result in 20. If more results than had been intended are found to fall outside both upper and lower control levels, this would indicate that the variation in the results is higher than expected and it may be necessary to tighten up the control to reduce the standard deviation or, if this does not seem to be practicable, to increase the average strength to compensate for the greater variation.

In this use of the control chart it will be seen that no calculations are required apart from those for the initial setting out of the control levels. Even these calculations can be eliminated if probability paper is available: a straight line is drawn through the intersections of the average strength and 50 per cent line with an appropriate slope to intersect the 16 and 84 per cent lines at $\bar{x} \pm \sigma$; the control levels for other proportions can then be read off the graph. On the other hand, this technique suffers from the disadvantage of any analysis based on the range of extreme values of a set of results in that the characteristics of variation are assessed from only a few results.

In order to make use of more of the available results and to observe trends in the average strength and in the variation in strength independently, control charts with appropriate control limits can be drawn for these properties of the results. The limits are calculated from a value for the standard deviation obtained from past experience and not from an estimate obtained from the results under examination, unless there is a sufficient number to give a more reliable estimate.

It is usual to prepare a control chart for the averages of the results in sets of a given number of results to show up any trends in the average strength of the concrete: the control limits are set at levels appropriate to a particular probability assuming the variation is random and having the standard deviation used in the calculation of the required overall average. If more than the expected number of values occur outside the levels, either a systematic trend is occurring in the average strength or the standard deviation assumed in fixing the control levels was too low, and it may be necessary to modify the concreting procedure. The levels differ from the required average by $k\sigma \div \sqrt{n}$, where k is the value associated with the probability, and taken from fig. 3, and n is the number of results in each set. Thus, following the method of specifying the quality of concrete proposed in chapter 4, where there are to be four results to a set, the control levels would be $1 \cdot 28\sigma \div \sqrt{4} = 0 \cdot 64\sigma$ above and below the average for a chance that one value in 10 would be above the upper level and 1 in 10 below the lower level, *i.e.* there would be 2 out of 10 values outside both levels. Similarly, the difference between the expected average and the control levels for 19 out of 20 values to occur between them would be $1 \cdot 96\sigma \div \sqrt{4} = 0 \cdot 98\sigma$. This is shown in fig. 13, illustrating the example given in Appendix 2.

The variability of the concrete can be examined for systematic variations by preparing control charts for either the range or the standard deviation of sets of results. It is obviously easier to use the range than the standard deviation because the amount of calculation involved

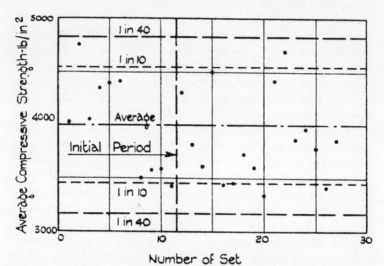

Fig. 13—Control chart for averages of 4 results

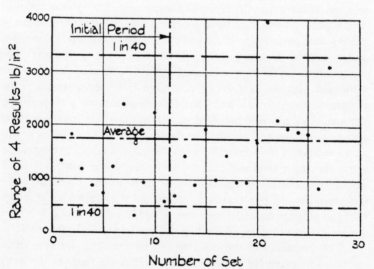

Fig. 14—Control chart for ranges of 4 results

is much less, but the result is less accurate, and especially so if the original results are not distributed approximately 'normally'; the number of results in the sets from which ranges are calculated should not exceed 11 or 12. It would seem that the preparation of control charts for the ranges of four strength results could be very helpful on many jobs.

The preparation of a control chart for the ranges of sets of four results involves the calculation of the expected average value of the range ($\bar{w}$) from $2 \cdot 059\sigma$ and of the control levels, based on 19 out of 20 values occurring between them,[47, 49, 50b] from $0 \cdot 59\sigma$ and $3 \cdot 98\sigma$; the corresponding factors[50a] for 9 out of 10 results are $0 \cdot 77\sigma$ and $3 \cdot 65\sigma$, respectively, and for 99 out of 100 results are $0 \cdot 38\sigma$ and $4 \cdot 65\sigma$. The factors for other numbers of results per set up to 12 may be obtained from B.S. 600 (both original and reprint).[50a, b] These levels are shown in fig. 14, illustrating the example given in Appendix 2.

From what has been said in chapter 4, there can be little accuracy expected from values of standard deviation calculated from small numbers of results; on the other hand, there is little likelihood of having large numbers of strength results available from concreting work. Control chart data for standard deviations is based on the actual standard deviation of the results in the set, s (where n, rather than n–1, is used in the denominator of the expression), and not on the estimate of the standard deviation of the consignment, $\sigma$. Hence for control purposes it would be convenient to have 10 results per set to eliminate complicated division from the calculation of the values of standard deviation. In this case the control levels would be fixed at $0 \cdot 923\sigma$ for the average value, $\bar{s}$, and at $0 \cdot 520\sigma$ and $1 \cdot 379\sigma$ for levels between which 19 out of 20 results would be expected to occur.[47, 49, 50] An example, illustrating the application to a job with over 400 results, is given in Appendix 2.

The factors by which the expected standard deviation is to be multiplied in calculating the control levels are obtained on the assumption that the results will occur completely at random. However, it was pointed out in chapter 4 that systematic trends, which cannot easily be controlled, are included in the variation throughout the job; these trends have the effect of making both the average of the ranges and of the standard deviations calculated from sets of results less than the values that would be obtained by calculation from all the results for the job. This effect was shown in chapter 4 in relation to the calculation of $\bar{x} - k\sigma$, and it can also be seen in the examples of control charts given in fig. 14 and Appendix 2 (fig. 19). The control charts for range and standard deviation therefore tend to give an optimistic picture of the control position, especially if there are small numbers of results per set, and remedial action should not be delayed if the charts indicate any tendency for the operation to be going out of control.

B.S. 2564: 1955[49] gives guidance on control chart technique particularly applicable to manufacturing to dimensional tolerances; the procedure to be adopted depends on the relative precision index (R.P.I.) defined, for our purpose, as the difference between the design average and the specified minimum divided by half the average range. It seems that for most economical concrete production the control would be classified as being of low relative precision: for four results per set the R.P.I. would be less than 3.0. If the R.P.I. for sets containing four results were found to be more than 3, the control would certainly be adequate if the average was in control.

If changes made in the concreting procedure are expected to cause significant changes in either the average strength or the variability of the strength we would expect that these effects would be shown as trends on the control charts. However, when working the other way round and examining the variations recorded on the control charts, we should not jump to conclusions too easily, nor try to attribute all the features to minor changes in the concreting operations, because many may be the result of random variations.

# METHODS OF CALCULATING THE STANDARD DEVIATION

According to its derivation, the standard deviation is the square root of the variance, which is defined in terms of the squares of the deviations (*i.e.* the differences between each individual value and the average value) of a set of results. The formula for the standard deviation, therefore, is often expressed in terms of the deviations as follows:

$$\sigma = \sqrt{\frac{\Sigma (x - \bar{x})^2}{n - 1}}$$

However, this is not a very convenient form for control purposes because the average has to be recalculated as the work progresses. A more convenient form which allows of simple progressive computation as the results become available is:

$$\sigma = \sqrt{\frac{\Sigma(x^2) - \dfrac{(\Sigma x)^2}{n}}{n-1}}$$

A calculating machine can be of considerable help in making these computations and, indeed, some machines are specially adapted for determining the sum of the numbers, $\Sigma x$, and the sum of the squares of the numbers, $\Sigma(x^2)$, at the same time.

Where large numbers of results are to be treated as one set, as at the end of a job, the computational work can be considerably reduced, without a great loss of accuracy in the estimate, by grouping the results into relatively narrow intervals, as is done in preparing a histogram. The numbers of results $n_1$, $n_2$, $n_3$, etc., occurring in each interval are then determined and are assumed to occur at the mid-values $x_1$, $x_2$, $x_3$, etc., of the intervals. The standard deviation is then estimated from the expression

$$\sqrt{\frac{\Sigma(n_1 x_1^2) - \dfrac{(\Sigma n_1 x_1)^2}{\Sigma n_1}}{(\Sigma n_1 - 1)}}$$

In making these calculations there can often be considerable simplification by changing the scale of the values. For example, in dealing with the compressive strength of concrete, the numbers to be treated mathematically are generally in the thousands and this involves several digits when summing large numbers of results. If all the results are reported to the nearest 50 lb/in², the first simplification would be to divide all the compressive strength results by 50 to give a new scale of values in integers. Thus 4,000 lb/in² would become 80, 4,050 lb/in² would become 81, 4,100 lb/in² would become 82, and so on. Next, it will be seen that the new scale of values still consists of relatively large numbers which could be reduced by taking an arbitrary zero much nearer the set of results themselves. For example, if all the results lay in the range 3,500 to 4,500 lb/in², it would be possible to take 3,500 lb/in² as an arbitrary zero and all the results would then be positive integers ranging from 0 (equivalent to 3,500 lb/in² or 70) to 20 (equivalent to 4,500 lb/in² or 90). The size of the numbers to be manipulated can be reduced even further, if they include both positive and negative values, by moving the arbitrary zero to a value near the average: in this case, if the zero were made to be equivalent to 4,000 lb/in², the numbers to be used in the calculations would range from −10 to +10.

The values of both the average and the standard deviation would then be determined on the new arbitrary scale and would have to be converted back to the original scale. The true value of the standard deviation would be obtained simply by multiplying the derived value by 50; to determine the true value of the average it would be necessary to correct for the transference of the zero from the true value to the arbitrary value and to multiply the result by 50. For example, if the average were found to be 10·7 on the arbitrary scale, with the zero set at 70 (equivalent to 3,500 lb/in²), the true average would be:

$$(70 + 10·7) \times 50$$
$$= 80·7 \times 50$$
$$= 4,035 \text{ lb/in}^2$$

Similarly, if the standard deviation were found to be 14·3 on the arbitrary scale the true value for the standard deviation would be:

$$14·3 \times 50$$
$$= 715 \text{ lb/in}^2$$

## STANDARDS OF SITE CONTROL

So far the variability of concrete strength has been discussed in terms of the standard deviation rather than of the coefficient of variation. This approach has been taken for the sake of simplicity, because an additional calculation is required to determine the coefficient of variation from the standard deviation and the average value, and because it is probably easier to assimilate the significance of the variation when expressed in the same units as the property under examination.

In the early days of the study of concrete variability many authors assumed that the different systems of controlling the quality of concrete could best be distinguished by expressing the variability as a coefficient of variation rather than as a standard deviation.[5, 6, 14, 18] This was no doubt a sufficiently close approximation for the relatively narrow range of average strengths and different types of control systems that were within the range of experience then being examined. However, it became obvious, as more interest was taken in high-strength concrete, that the use of the coefficient of variation appropriate to a particular system of batching and mixing concrete, derived from concrete with an average strength of about 3,000 to 4,000 lb/in$^2$, would not be appropriate if the average strength were of the order of 6,000 to 8,000 lb/in$^2$.

For example, if concrete had an average compressive strength of 4,000 lb/in$^2$ and the coefficient of variation was 12$\frac{1}{2}$ per cent, the standard deviation would be 500 lb/in$^2$. We would expect that 19 results out of 20 would fall within a range of about:

$$\pm\ 2 \times 500 \text{ lb/in}^2$$
$$= \pm\ 1,000 \text{ lb/in}^2$$

*i.e.* the concrete strengths would range from about 3,000 to 5,000 lb/in$^2$. If the same coefficient of variation applied also to concrete with an average compressive strength of 8,000 lb/in$^2$, the corresponding range of strength results to be expected would be 6,000 to 10,000 lb/in$^2$.

This range for the higher strength concrete seemed too wide for such a high level of control at the lower strength, and some people[8, 15, 40, 42] therefore suggested that the standard deviation was the more appropriate to represent a particular standard of control. Those who favoured the coefficient of variation as being the better assessment of the standard of quality control were then quick to point out that the standard deviation was equally unreliable when dealing with low-strength concrete. If, in the above example, the standard deviation of 500 lb/in$^2$ were to be taken to characterise the system of control, the corresponding range of strength results likely to be obtained if the average strength were to be 1,000 lb/in$^2$ would be from 0 to 2,000 lb/in$^2$ —which again did not agree with experience.

Both experimental[7] and theoretical[51] approaches to the problem have shown that the coefficient of variation more nearly represents a particular standard of control at relatively low strengths, while the standard deviation more nearly represents the standard at high strengths. Such a compromise, however, is not very satisfactory when numerical information is required about the variability of concrete strength appropriate to a particular system of control. As the compressive strength of any particular type of concrete is closely related to the water/cement ratio of the mix, and the variability in the strength of the concrete made on site is probably related directly to the variability of the water/cement ratio, it seemed appropriate to investigate whether any characteristic of the variation in water/cement ratio could account for the observed variation in strength.

It has been found[43] that a particular standard of quality control can be associated with a 'control ratio', defined as:

$$\frac{\text{the water/cement ratio required to produce the mean strength}}{\text{the water/cement ratio required to produce the minimum strength}}$$

where the minimum strength has a statistical significance. Table 5 gives values of the control ratio appropriate to different standards of quality control where the minimum strength is that value below which 1 in 25, 40 or 100 individual results would be expected to fall. Alternatively, fig. 9 can be used to determine the standard deviation appropriate to the same standards of control for different average compressive strengths; the total range of the vertical scale has to be adjusted according to the degree of control, as indicated in the legend to that figure.

## TABLE 5
### Control ratios for different standards of control

Control A: Batching cement and aggregates by weight with servo-operation
Control B: Batching cement and aggregates by weight with manual-operation
Control C: Batching cement by weight and aggregates by volume
Control D: Batching cement and aggregates by volume (not continuous mixer)

| Proportion of cube results expected below minimum | Supervision | Control ratio | | | |
|---|---|---|---|---|---|
| | | Standard of control | | | |
| | | A | B | C | D |
| 1 in 25 | Poor | 0·78 | 0·75 | 0·72 | 0·70 |
| | Normal | 0·82 | 0·79 | 0·77 | 0·75 |
| | Good | 0·86 | 0·83 | 0·82 | 0·80 |
| 1 in 40 | Poor | 0·76 | 0·71 | 0·69 | 0·66 |
| | Normal | 0·80 | 0·76 | 0·74 | 0·72 |
| | Good | 0·84 | 0·81 | 0·79 | 0·78 |
| 1 in 100 | Poor | 0·71 | 0·67 | 0·63 | 0·60 |
| | Normal | 0·76 | 0·72 | 0·69 | 0·67 |
| | Good | 0·81 | 0·77 | 0·75 | 0·74 |

It seems from the British survey of the variability of concrete made on many sites[43] that the type of equipment employed for controlling the quality of the concrete affects the variability to a limited extent only and that other unaccounted causes of variation have a considerable effect such that there is wide overlapping of the variability from one standard of control to another; this unknown factor has been attributed to the amount of supervision exercised on the site. It seems, therefore, that having expensive equipment does not necessarily guarantee a high degree of control and that a small variation in concrete strength can be obtained with less expensive equipment if adequate supervision is exercised.

# CHAPTER SIX

# Constituent Materials
# of Concrete

## TYPES OF SPECIFICATION REQUIREMENT

Cement and aggregate, the main constituents of concrete, are usually bought and sold subject to the requirements of a recognised specification. The limitations imposed on the quality of these materials are intended to ensure that they are satisfactory for making concrete. Sometimes, as in the case of aggregates, it is not always possible to express the requirements in terms of limiting values for the results of tests made on samples of the material, and reliance has then to be placed on a description of the characteristics required. But, wherever practicable, the required property has to be checked by making appropriate tests and rigid limits are applied to the result obtained; there is no intention that these maximum or minimum values should carry any statistical significance. A consignment of the material is liable to be condemned, therefore, if the results of tests on samples taken from that consignment do not come within the appropriate limits.

This tradition of having rigid limits in the specification is no doubt convenient in its simplicity, but when applied to some of the properties it does bring with it problems like those found in specifying the minimum strength of concrete. For example, Portland cement specifications usually require that the strength-making characteristics shall be checked by testing specimens after they have been cured for three or seven days. On the face of it, therefore, it would seem that the cement manufacturer must treat his product in batches, which are sampled and tested and then set aside separately until the results of the tests become available; the batches would then be released or condemned according to the results of the tests. As this is not a practical procedure, the manufacturer is much more likely to arrange the control of the production of his material so that the average strength is high enough in relation to the specified minimum value for the risk of failure of a consignment to be so small that he is willing to stake his reputation on the fact that all the cement delivered will be of satisfactory quality. He therefore cannot be sure that every consignment of cement that is sold will have a strength above the minimum required by the standard, but he will be prepared to guarantee the material and make good any consignment found to be unsatisfactory.

The properties which are specified fall broadly into two categories: those which have to be controlled so that the durability of the concrete will not be adversely affected, whatever the type of concrete manufactured, and those which have to be controlled because they have a direct effect on important characteristics of the concrete made from those materials. Examples of the first type include the soundness and some of the limitations of the chemical composition of cement and the limitation of the silt content of fine aggregate; examples of the second are the strength-making properties of the cement and the grading and particle shape characteristics of the aggregate. For most site work the difference between the two categories is probably not very important because the amount of regular testing to determine the second type of property is usually too small for the statistical analysis of the results to be of real value. However, where a contractor is interested in applying statistical techniques to the control of the quality of his concrete he could make good use of any records of this type when choosing the materials to be used. If an aggregate supplier, for example, could show him records of routine works tests, which indicate the range of such properties as the grading and particle shape of the material, he would be in a better position than his competitors who could not produce this evidence, even though their material may equally well comply with the requirements of the standard.

## STRENGTH OF CEMENT

The manufacture of cement is usually so well controlled that there is little doubt that the material will comply with the requirements of the standard. This being so, one might expect that contractors buying cement would take as little interest in its quality as they do in the quality of the reinforcing steel; however, this is not so. The steel is a final product, whereas the cement is a raw material for a further process and the contractor often takes a considerable interest in the strength of the cement because it affects the way in which he makes satisfactory concrete.

The contractor's interest in the strength of the cement is not so much in whether the cement complies with the specified minimum strength—that he takes for granted—but by how much the strength of the particular cement he receives exceeds the minimum value. He knows that the actual strength of the cement in a consignment has a direct and important effect on the actual strength of a particular type of concrete made with that cement; otherwise there would be no point in specifying the strength of cement in terms of a test involving the making of concrete cubes with samples taken from the consignment.[52] With this in mind, some contractors have gone so far as to request that the cement delivered to them (at the normal price) should all be above average quality, so that they will be able to produce concrete complying with a required minimum compressive strength with more economical mix proportions. Other contractors have complained when they have found that they have received cement which is below average. This type of comment obviously arises because of a lack of appreciation of what is meant by the term 'average', as already defined in the first chapter. Obviously, about half of the cement that is produced at a works must be above the average strength for that works and half below the average strength so long as the variations occur at random; unless deliveries are selected, there is a 50/50 chance that the contractor will obtain cement either below or above average.

One way to help to bring this idea home is to imagine that we are one of a random group of people each of whom knows his own intelligence quotient; the average I.Q. for that group could therefore be calculated. We need not be unduly ashamed if our own I.Q. is below the average for the group because there will be about half the people in the group who fall into the same category; in any case, this sort of statement does not give any indication of whether the average or the minimum I.Q. is high or low.

These variations in the strength of cement above and below the average value are normal and are only to be expected. A contractor would only be justified in complaining if he found that the strength of the cement delivered to him was below the specified minimum value.

It may be that some of the anxiety over the difference in the strength of cement supplied to a job arises because we do not distinguish between the effect of an individual consignment of cement on the actual strength of the concrete made with that consignment from the effect of the variation in the strength of different consignments of cement on the variation of the strength of the concrete made from them. This distinction is important.

## ECONOMICAL MIX DESIGN

A contractor, having to design a concrete mix to comply with a specified minimum strength requirement, must, like the cement manufacturer, set his target of production at an average value appreciably above the minimum value specified to allow for the variation which will occur in the quality of the concrete due to many factors, of which the strength of the cement is only one. If he is so concerned about the effect of the variation in the characteristics of the cement on the strength of the concrete he could design the mix on the assumption that he would receive only cement of the specified minimum strength and make due allowance for all the other causes of variation in concrete strength when fixing the level of the average strength he had to aim for. This procedure, however, would be unrealistic because, to be consistent, he should treat all the other variables which will affect the strength of the concrete in the same way as he treats the cement, and he would then find that the mix would be so uneconomical as to be impracticable.

The more sensible approach is to regard differences in the properties of the cement as being one of the normal variables, along with the many other causes of variation, and to include them all together in the design of the mix. It then becomes apparent that the effect of the variation in cement strength upon the variation in concrete strength is very much less significant than the effect of the strength of an individual consignment of cement on the strength of a single batch of concrete made with that cement.

Statistical examination of the relative importance of the variation in the cement quality, and of the various other factors which contribute to the variation in the strength of concrete taken all together, has shown that the proportion of the total variation due to the variation in the characteristics of the cement is quite small unless all the other causes of variation are very carefully controlled.[8, 15, 43]

The effects of cement variation with varying degrees of site control are given in table 6; it shows the amount by which the variation in concrete strength could be reduced if it were possible to use cement which was always of exactly the same quality throughout the job. As there is no hope of ever achieving this in practice, it also shows how small is the effect of a more practical reduction in the variation of cement quality on the variation in concrete strength. The effect of drawing cement from only one works instead of from many works (which means the unlikely situation of having cement from most of the works in the country) will be to reduce the standard deviation of compressive strength by not more than 10 per cent when the highest standard of control is used. Hence there is little point in trying to obtain cement of limited variation except perhaps for the most highly controlled jobs.

### TABLE 6

**Percentage reduction in standard deviation of concrete strength due to removal of different sources of variation**

| Source of variation | Percentage reduction in standard deviation | | | |
| --- | --- | --- | --- | --- |
| | Standard of control (see Table 5) | | | |
| | A | B | C | D |
| Removal of variation due to cement (drawn from many works) .. | 32 | 23 | 19 | 16 |
| Removal of variation due to cement (drawn from one works) .. | 26 | 18 | 14 | 12 |
| Reduction of variation due to cement being drawn from one works instead of many .. .. | 8 | 6 | 5 | 4 |
| Reduction in inherent variation of concrete strength (i.e. not including the testing error included in observed results) due to cement being drawn from one works instead of many .. .. | 10 | 7 | 6 | 5 |

Further, we must again distinguish between the effect, on the design of the concrete mix, of the variation in the quality of cement and of the average value of the strength-making characteristics of consignments of cement. If a large number of measurements of strength at a particular age were made at regular intervals on supplies of ordinary Portland cement from many works it would be found that the compressive strength ranged from somewhere near the specified minimum to a value as much above the average compressive strength for the consignments as the minimum is below the average.

If, however, the consignments of cement delivered from particular works were examined at frequent intervals, not only would we expect the variation to be less, but the average and the standard deviation of the results might vary appreciably from works to works. From this information it would be possible to assess whether some sources of supply were more suitable for highly controlled work than others. A number of the larger contractors, with appropriate facilities for regular testing, have found this to be a valuable exercise and have used their data in arranging for the delivery of cement to particular jobs.

## AGGREGATE QUALITY

The supplier of aggregate, which is claimed to comply with B.S. 882 : 1954, as amended in 1956,[54] must satisfy himself that the output at the source of production consistently conforms to the requirements of the appropriate specification of the standard. He is in a similar position to the cement manufacturer in that it would be quite impracticable for him to test samples from every load of material delivered and not release those loads until the results were available. But by sufficient routine testing and examination of the results by statistical methods he could control his output at the source of production, with an adequate margin of safety against failure to comply with the requirements of the standard, so that he could guarantee to make good any consignments found to be non-standard. As with the control of the quality of concrete, the frequency of taking and testing samples of aggregate would be greater at the start of a statistical examination of the variation in production, so that an indication of the level of control can be obtained relatively quickly, than subsequently, when the testing is undertaken to check that a reasonable degree of control is being maintained.

One curious feature of some concrete specifications is the demand for what might be termed a 'reference sample' of aggregate. Some specifying authorities require that the supplier of aggregates must submit a sample of each of the materials for the approval of their representative, and that these materials will then be regarded as the reference samples by which any further supplies will be judged. This places the supplier in a difficult position. Should he send a sample of his best material and run the risk that all other samples taken from the consignments delivered will fall short of it and be condemned? Should he send a sample of his worst material, knowing that none of his material is likely to be condemned, in the very unlikely event of his material being accepted in competition with that of other suppliers? Or should he, as is sometimes required, submit what might be called a 'typical sample' with the possibility that some of his consignments may still not be up to standard?

This practice is not consistent with the idea that a specification is a document having the express purpose of making clear the intention of one party to the contract to the other. It is rare to find in a specification the properties of the aggregate that are to be associated with the so-called 'reference sample', e.g. whether it is the grading, the particle shape, the colour, the modulus of elasticity, the absorption, the durability, the silt content or any other property which may be required to be not worse than that in the given sample. If an accepted standard for the material is not adequate, e.g. if B.S. 882 does not safeguard the suitability of the aggregates for a special purpose, the specifying authority ought at least to say in what respects the aggregate must be superior; the supplier will then know whether he is likely to be able to comply with the requirements and not just be at the mercy of the whims of the specifying authority's representative.

Although the overall grading of an aggregate has an important effect on the economy of producing a concrete mix with a certain minimum strength, the effect is not so direct as that of the strength of the cement. Coarse and fine aggregates having a considerable range of grading can be used equally well to produce concrete of the same workability and strength, using the same richness of mix, provided they are used in relative proportions which produce the desired overall grading; variations in grading of the separate materials without a compensating change in relative proportions will, however, affect the quality of the concrete.

When a contractor has a choice of sources of supply of aggregate, therefore, he may want to know the average grading of the various sizes of aggregate being offered so that he can assess how readily the materials can be combined to produce an economical mix. But, in addition, when he is controlling the quality of the concrete, the range of variation in grading, likely to occur from one consignment to another, also becomes important, because he would not want to keep changing the relative proportions of the different sizes of material to accommodate the fluctuations in grading occurring throughout the course of the job. If, therefore, the aggregate supplier can produce records of routine testing at the works which indicate a small range of grading, this information, which is not called for in the standard, might well be as useful to the contractor as a certificate stating that the material was guaranteed to comply with the grading requirements of the standard.

The particle shape and texture is a characteristic of the aggregate which is more akin to the strength of the cement in that it affects the quality of the concrete directly. For example, a richer mix is normally required to produce concrete having a certain workability and compressive strength when the aggregate particles are angular and rough than when they are rounded and smooth, particularly in the fine aggregate range. A varying proportion of crushed material in a natural sand might, therefore, have a significant effect on the variation in the concrete quality, and routine control testing might be a worthwhile precaution against obtaining excessive proportions of crushed material.

The moisture content of aggregates at the time of batching is an important factor in the control of concrete quality. However, the moisture content at the time of delivery may also be of significance in affecting the amount of material purchased, especially where the aggregate is sold by weight, and, to a limited extent, where fine aggregate is sold by volume and may be subjected to variable bulking according to the variation in moisture content. A statistical examination of the moisture content of each grade of aggregate at the time of delivery could lead to interesting results.

## TRIAL MIXES

Before starting concreting operations a contractor may well wish to make laboratory or full-scale trial mixes to determine whether the mix proportions he has chosen for a job will be satisfactory. Even when the trial mixes are made under the relatively closely controlled conditions of a laboratory, the results of the tests may be misleading unless the contractor knows how the samples of material used in making the trial mixes are related to the consignments of materials he will be receiving on the job. It is advisable, therefore, that the strength of the cement and such important characteristics of the aggregate as the grading and particle shape and texture should be measured in independent tests on samples of the materials to be used in the trial mixes. This information is of little value by itself, but if records of the variation in the corresponding properties of the materials over a long period of time are available we can assess whether the compressive strength of the trial mix is likely to be above or below the value that would have been obtained if the samples of materials had been at their average values. Even if this information is not available from the suppliers of the materials, a knowledge of the properties of the samples used in trial mixes could be of some help if an investigation of the causes of site test results being unexpectedly high or low became necessary later.

If a contractor has a choice of different sorts of aggregate for a particular job, he may wish to make trial mixes with samples of the materials before he decides which he will use. Again, if the results of these tests are to be of any real value, considerable care would have to be taken to ensure that the samples used in the trial mixes were truly representative of the average run of production of each source of supply; or, alternatively, the properties of the aggregates would have to be measured and, as far as possible, related to the variation in these characteristics as determined from routine control testing undertaken by the supplier. Otherwise the differences in the results obtained from the trial mixes would have to be quite large before it would be possible to conclude with any certainty that one material was likely to be more suitable than the other.

Trial mixes may also be made on occasions to check whether the use of an admixture would lead to more economical mix proportions for a certain job. For this purpose there would probably be less need to ensure that the samples of cement and aggregate were truly representative of that job, but sufficient care would have to be taken to ensure their uniformity from one batch made with the admixture to the next made without it.

In each of these examples the tests should be made under conditions which are controlled as closely as possible to eliminate as much of the unwanted testing error as is practicable. This, in effect, means following the detailed procedure laid down for the making of trial mixes in appropriate standards. Even after taking all these precautions there will still be some residual testing error which may mask the true differences which we are seeking to establish. Part of this testing error will occur within batches and part of it between batches. We are not normally interested in the error occurring within the batches because we can normally make an adequate number of tests on the batch to obtain a fairly accurate estimate of the average value for the batch. For example, it is usual to make at least three specimens for measurement of strength at each age and, similarly, we would expect to make at least three determinations of workability on each batch.

When trial mixes are made to compare different materials, such as aggregates from two different sources or mixes with and without an admixture, statistical calculations can be made to give some indication of the likelihood of the difference in the average values being significant.[55] The procedure for making these calculations is given in Appendix 1.

# CHAPTER SEVEN

# Finished Concrete

## EXAMINATION OF SUSPECT CONCRETE

So far we have been mainly concerned with the drafting of specifications governing the strength of concrete and with the methods of producing concrete to comply with their requirements; in all this work we have to rely on our test procedures. But the taking of samples, the making of test specimens and the measurement of strength are really only a means to an end and not the end itself: the ultimate requirement is that the concrete in the structure should be satisfactory. If, therefore, our sampling and testing procedures are at fault, we may sometimes have what appears to be unacceptable concrete due to errors in our testing method, and further examination of the structure itself may be thought desirable; rarely do mistakes in the making of test specimens lead to unduly high compressive strength results.

Some types of finished concrete can be checked readily. For example, certain precast units can be examined by testing samples taken from a consignment; cylindrical cores can be extracted from road slabs; and sections of *in situ* concrete, such as columns and beams, can be tested non-destructively.[56, 57] Load testing of structures may occasionally be necessary. In the former cases the concrete will be subjected to a test which is not the compressive strength test, on which the specification is based, but which is thought to correlate to some extent with that test. It would be unwise to assume that a precise relationship, determined generally from some research work, would be appropriate for the particular example in question, especially when the relationship is likely to be affected by the type of materials used and differences in the methods of compacting and curing the concrete in the structure and in the test specimens. It is advisable, therefore, that the procedure to be adopted for investigating any defective concrete should be applied also to samples of acceptable concrete in the structure, as assessed by the compressive strength test.

As many samples as economically possible should be obtained for examination, including samples of material which is thought to be little better than the minimum acceptable. Subsequent examination of the samples will then show whether the variation in quality of both the acceptable and the suspect material is low enough to show a difference sufficiently significant for justifying the rejection of the suspect material.

The principle may be illustrated by the following example involving the use of non-destructive testing as a means of checking whether hardened concrete is likely to have been up to the required standard.[58] The concrete is required to have a certain minimum compressive strength; the result of a test shows that this requirement has not been met, but there is some doubt whether the failure to comply with the requirement is due to a lack of quality of the concrete or due to some mishap in the testing procedure. Results obtained from the non-destructive testing method are known to correlate reasonably well with the compressive strength for any one particular type of concrete, but the correlation is not known for this material.

The technique to be adopted should be as follows. A number of examples of concrete corresponding to satisfactory compressive strength results are located and for each of these a set of several non-destructive test results is obtained; a similar set of non-destructive tests is made also on the concrete corresponding to the low strength result. If the non-destructive results are plotted graphically against the corresponding compressive strength results, as shown in fig. 15, we can see how the relationship between the results of the non-destructive and strength tests for the concrete with the low strength compares with that for the acceptable material. If the results appear to be consistent, as in fig. 15, rejection would be justified, but if the variation in non-destructive test results for the suspect concrete cannot be distinguished from that of the acceptable concrete, as in fig. 16, the low strength was probably due to some other factor than poor quality of the concrete itself, and rejection would not be justified.

The decision to accept or reject may be difficult in borderline cases. Statistical calculations may be of some help if a suitable correlation curve between the two properties can be drawn through the points for the acceptable concrete and extrapolated to lower strengths, ignoring the results for the suspect concrete, but drawing on other experience if it is available. The scatter of the non-destructive test results about the value at the intersection of the curve and the corresponding compressive strength ordinate is then treated as described in the section on trial mixes in chapter 6 and Appendix 1, to determine whether the suspect

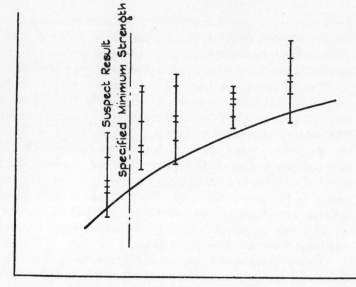

**Fig. 15**—Suspect concrete probably below standard

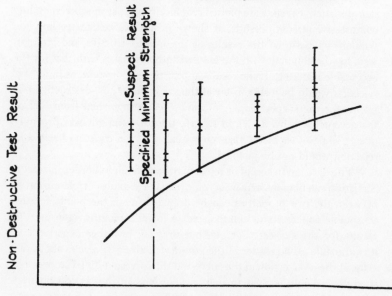

**Fig. 16**—Suspect concrete probably satisfactory

results show a significant departure from the remainder of the results. The deviations of the acceptable results are treated together and the deviation for the suspect concrete compared with them.

When we seek to examine a sample of finished concrete by some test which is thought to correlate with the standard compressive strength test we need to look at the correlation from two points of view. Firstly, the new test should have a small testing error because any correlation of one set of results with the other will tend to be masked by the testing errors of both methods of test. Secondly, we must examine how readily the new test discriminates between different types of concrete, as classified by the original test.

For example, the velocity of an ultrasonic pulse through hardened concrete correlates broadly with the compressive strength of that concrete, although it is also affected by other properties. If we were to examine the values of pulse velocity through a number of appropriately shaped specimens of similar concrete we might consider the variation in the values to be small and be tempted to think that the testing error was therefore small enough for this method to correlate usefully with compressive strength. However, further examination may show that the range of pulse velocity for the range of compressive strength in which we were interested was also quite small and that the usefulness of this test in discriminating between different classes of concrete in terms of their compressive strength with any degree of certainty was somewhat limited unless a very large number of measurements was taken. The same thing may be found with the rebound-hammer method of non-destructive testing. The usefulness of both these non-destructive tests, in discriminating between concretes of different compressive strength, is greatest for low-strength concrete and diminishes as the compressive strength increases.

## PRECAST CONCRETE PRODUCTS

Many precast products are sold subject to a specification which requires the testing of the finished article rather than of the concrete used in manufacturing that article. The statistical control methods already described for the compressive strength specimens could equally well be applied to these acceptance tests, and it is sometimes useful to record also other data from the manufacturing process (e.g. the compressive strength of the concrete) which would be likely to correlate with the final result. A particular example of this sort of technique applied to the manufacture of precast prestressed floor joists has been reported.[59]

A carefully planned system of random sampling and testing of the cement, aggregates, concrete and finished units and of checking steel position and slip and curing conditions (carried out daily except for the cement testing) led to a constant watch being maintained on all vital stages of production. Confidence limits on the control charts were abandoned after a while because the control was so high: for example, the standard deviation of 124 pairs of one-day cube results obtained from concrete made with high alumina cement over a period of six months was a little over 300 lb/in² with an average strength of 9,113 lb/in², giving a coefficient of variation of under $3\frac{1}{2}$ per cent. The system was considered to be well worth while because it quickly revealed any tendency to deterioration in quality and generally indicated the cause of the trouble. The experience gained with the control of the smaller units gave confidence in manufacturing larger units which would have been much more expensive to test.

Mention was made in chapter 6 of the undesirable requirement in specifications that a supplier of aggregate should have to offer 'reference' samples. However, this type of clause, suitably modified, might sometimes be helpful when specifying the quality of precast products if there is some difficulty in defining the requirements. As an example, when units are required to have particular characteristics of surface finish, it would be reasonable to specify that, after a contract had been let, a number of these units be cast as in normal production and inspected; then one or more of the units would be selected, by agreement between the producer and the specifying authority, as a reference sample for particular properties, such as degree of compaction, surface imperfections, quality of arrises or colour. Any units produced subsequently which did not come up to these standards, set by the reference samples and which are now clearly defined and understood between both parties, could then be rejected. There is no doubt that these samples would not be 'typical' samples, but would be of limiting acceptable quality.

## CONCRETE BRICKS AND BLOCKS

The British Standards for concrete bricks and blocks (B.S. 1180[60] and B.S. 2028[61]) include requirements for the compressive strength of the units; this is reasonable because the type of concrete used for their manufacture is not easily made into standard test specimens. Originally the requirement was that a sample of 12 of these units should be tested for compressive strength and that the average value for the 12 specimens should not be less than a stated figure; indeed, this is still the only stipulation in the standard for concrete bricks.

This type of specification does not induce any real control over the variability of the units because it would be possible for some of them to have no strength at all provided others had more than adequate strength. Yet control of the variation in strength is probably of more importance for these units than for other types of concrete because at the present time the Code of Practice dealing with load-bearing walls (CP 111 (1948)[62]) bases permissible loads on the average strength of the units and not on the minimum strength as in most other design practice.

A revision of the standard for concrete blocks then included a requirement that the strength of the weakest block in the set of 12 should not be below 75 per cent of the average strength for the 12. This naturally had the effect of preventing the acceptance of very weak units and limiting the variability of the strength, but it brought with it two disadvantages: one, that there were now two requirements with which the blocks had to comply; and the other, that the acceptance or rejection of a consignment could depend on the strength of one block only. The specification for sandlime bricks (B.S. 187 : 1955[63]) overcame the difficulty of rejection on the basis of a single result by requiring, not only a minimum value for the average strength of a set of 12 bricks, but also that the average strength of the seven weakest bricks in the set of 12 should not be less than either 80 or 90 per cent of the average for all the 12: the percentage depended on the class of bricks. Thus there was control of the variability of strength, but again two requirements had to be met.

An amendment to B.S. 2028 in November 1959 withdrew the requirement that the strength of the weakest block in a set of 12 should not be less than 75 per cent of the average strength of the 12, but retained a requirement that it should not be less than 75 per cent of the specified minimum value for the average (e.g. for lightweight aggregate blocks the average should not be less than 400 lb/in$^2$ and the lowest result not less than 300 lb/in$^2$). This specification of minimum values for both the average and the lowest strength of a set of 12 does not effectively control the variability of the units because blocks with a high standard deviation can be made to comply with the requirements pro-

vided their average strength also is high. The effect of the amendment
on the chance of a consignment containing blocks of low strength can be
illustrated by use of the statistical methods already described, assuming
that the sample of 12 blocks was selected at random.

For simplicity we may regard a consignment of lightweight
aggregate blocks as just complying with the strength requirements of the
existing standard if the average strength of the consignment is 400
lb/in$^2$ and the statistical minimum for 12 blocks, *i.e.* $\bar{x} - 1 \cdot 38\sigma$, is taken
as 300 lb/in$^2$; this assumption means that there is a 50/50 chance of the
average and of the calculated minimum for the set of 12 blocks failing
to comply with the requirements. We can then calculate the chance of
blocks in the consignment having strengths below various other levels
as indicated in fig. 17 by the heavy line. Similarly, the chance of
failure is shown for other conditions of higher control where the average
strength is still 400 lb/in$^2$, and for other conditions where $\bar{x} - 1 \cdot 38\sigma$ is
still 300 lb/in$^2$, but the average is raised to 500 and 600 lb/in$^2$ to accommodate greater variability, as is permitted by the standard at the present
time. It will be seen that, because there is no upper limit on the average
strength, a high proportion of blocks could theoretically have very low
strengths. The dotted lines in fig. 17 show that prior to the issue
of the amendment (when the strength of the weakest block should not
be less than 75 per cent of the average strength for the 12 blocks)
there was a very large measure of protection against blocks having
unduly low strengths.

It would seem that a specification for compressive strength in
statistical form would be well suited to the control of this type of unit.
The calculated value of the minimum strength, in the form $\bar{x} - k\sigma$,
would make use of all the results for assessing both the average strength
and the variation and, at the same time, would amount to only one
requirement which had to be met. The specification of minimum
strength in terms of a statistical minimum value has already been shown
to encourage better control, because the average strength can be
reduced the lower is the variation in the product, and this brings with
it a greater effective margin of safety.

Principles affecting the operation of a statistical specification have
already been discussed in chapter 4, but the experimental data is not
necessarily relevant to the testing of precast concrete blocks. For rapid
calculation it is convenient if the minimum strength specified is a round
number; for lightweight concrete blocks the value of 300 lb/in$^2$ would be

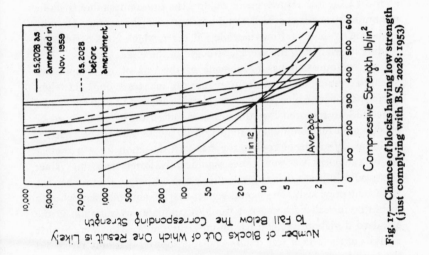

**Fig. 17**—Chance of blocks having low strength (just complying with B.S. 2028 : 1953)

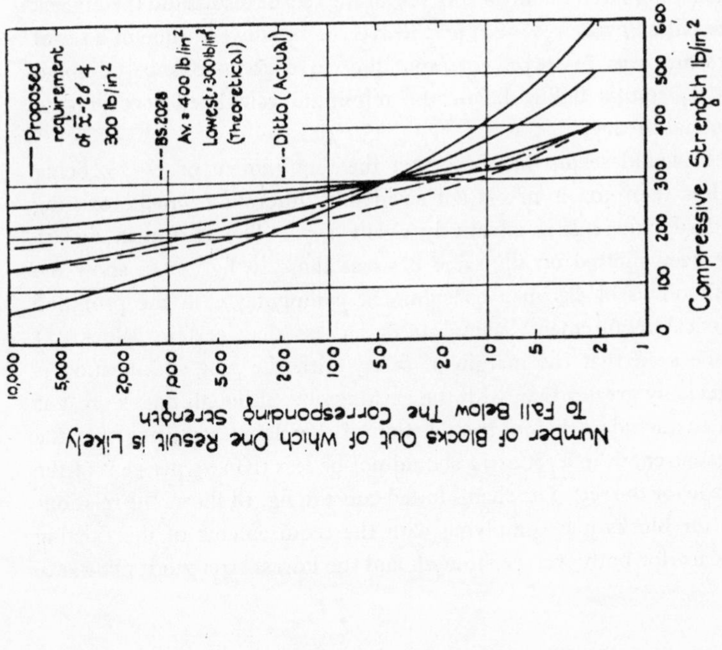

**Fig. 18**—Chance of blocks having low strength (just complying with proposed requirement)

convenient because it corresponds with the existing limit for the weakest of the 12 blocks. However, it would also appear desirable that the minimum strength specified should tie up with the average strength envisaged by those drafting the code CP 111[62], which gives permissible loading in relation to the average strength of the units.

An examination has been made of the sets of 12 compressive strength results obtained from many routine tests on lightweight aggregate concrete blocks and a considerable amount of data on aerated concrete. It appeared that the lowest standard deviation likely to be encountered for continuous production of blocks was of the order of 50 lb/in², although, of course, the standard deviation for a particular set of 12 blocks was sometimes significantly lower than this value; this would indicate that the value of k should be 2. From what has been said previously, one would expect that a value of k as high as 2 would be unrealistic because it would accentuate any variation in the standard deviation calculated from only 12 results. However, further analysis of the sets of test results available showed that only rarely did the lowest value in a set of 12 occur above the statistically calculated minimum corresponding to 12, *i.e.* $\bar{x} - 1 \cdot 38\sigma$. This did not appear to be due to an asymmetrical distribution of the results about the average, because, similarly, in only a very few of the sets of results did the highest value fall below $\bar{x} + 1 \cdot 38\sigma$; on the average the lowest value of a set of 12 results was $\bar{x} - 1 \cdot 7\sigma$, corresponding to a 50/50 chance of about 1 in 24 results falling below the minimum calculated according to statistical theory.

It would seem, therefore, that the requirement of $\bar{x} - 2\sigma$, being not less than 300 lb/in², is not so greatly different from the existing standard, where k is effectively about $1 \cdot 7$. The curves in fig. 18 have been plotted on the same basis as those in fig. 17 to show the likelihood of blocks in a consignment complying with the proposed statistical requirement having strengths less than certain values. It will be seen that the margin of safety with the new specification is appreciably greater than with the existing one, although not as great as with the standard before the deletion of the clause requiring that the lowest strength in a set of 12 should not be less than 75 per cent of the average for the set. The chain-dotted curve in fig. 18 shows the relationship for blocks just complying with the requirements of the existing standard for both average strength and the lowest strength if allowance is made for the difference between the actual and theoretical distribution of points.

Reference was made in chapter 4 to the variation in the estimates of the average and standard deviation of the consignment based on the test results of a sample and to the fact that the estimate of the standard deviation based on small numbers of results tended to lead to values which tailed off more to higher values than to lower values. Because the value of k proposed tends to exaggerate this effect, the manufacturer is more likely to be unfairly penalised than is the user to have to accept unsatisfactory material. It would seem reasonable, therefore, that the specification, which would otherwise condemn a consignment failing to comply with the requirement, might permit re-testing when the standard deviation observed exceeded a stated value, such as 150 lb/in², and compliance would then be checked by a calculation of $\bar{x} - 2\sigma$ from the 24 results available.

The proposals outlined above are being considered by the committee responsible for B.S. 2028 which is at present under revision.

## DIMENSIONAL TOLERANCES

With an increasing tendency for prefabrication in building and, associated with this trend, the greater need for modular co-ordination in building, control of the overall dimensions of precast concrete units is becoming of greater significance. At the present time it is recognised that any component in a building should be smaller than the space allotted to it; this means that all tolerances on the nominal dimension must be negative. The statistical control methods already outlined for compressive strength of concrete may be adapted in principle to the control of critical dimensions of those precast concrete units which are made in considerable quantities.

As we have already seen, there is no such thing as an absolute minimum strength and, similarly, we may find difficulty in manufacturing to limits on the basis of an absolute maximum dimension; for example, this could lead to excessively wide joints which might be undesirable for a number of reasons. However, it would be surprising to find specifications of dimensions with statistical limits attached to them because there is not likely to be any danger associated with the manufacture of an oversize unit—the penalty will be the probable rejection of the unit. Manufacturers would, therefore, have to set their own standards of control to produce an economical solution to all aspects of the problem.

It may be thought that the dimensions of concrete blocks would not be critical because any reasonable variation could be taken up in the width of the mortar joint. However, this is not necessarily true if the blocks are being used for an in-filling panel in a framed building which has been designed on a modular system to co-ordinate with the nominal size of the blocks. Any systematic trend away from the proper dimension, which allows for a ⅜-in joint, may lead to considerable difficulties for the blocklayer. A departure in average length will affect the spacing of the blocks in a course between columns and a departure in average height will affect the thickness of the bedding joint to maintain the correct number of courses between floors. On the other hand, the blocklayer may well be able to cope with reasonable variation in dimension from the appropriate true dimension provided he has a random selection of the blocks, including some oversized and some undersized, in any consecutive operation.

The manufacture of concrete blocks is such that the length and width are governed by the dimensions of the mould and are subject to appreciably smaller variations than the height. The blocklayer is therefore less likely to have difficulty in fitting the blocks into a particular course than in fitting the courses between the floors, particularly if the height of the blocks is subjected to a systematic error.

Precast floor beams may be considered as an example of mass-produced precast concrete units whose dimensions are important. Any variation in camber will lead to a step between adjacent beams at both the upper and lower surfaces, which, if excessive, can lead to extra expense in laying the floor screed or plastering the under-surface. Control testing of this characteristic would enable the manufacturer to decide whether there would be any overall economy in selecting joists according to their camber. The overall width of the cuboid space occupied by a floor joist after any distortion has occurred is also important. A statistical examination of the results of tests to measure this effective width might well be of considerable advantage to the prospective user.

The tale is told of a disgruntled contractor who ordered, let us say, 100 special precast concrete floor joists, 1 ft wide, for a building 100 ft long. They were delivered to the site and duly paid for. When the contractor started to erect them all went well until the end of the job, when he found that he only needed 99 because their average effective width was greater than 1 ft. Naturally, he complained to the manufacturer and asked him to take back the extra beam and refund him his money. This the manufacturer refused to do because the average actual width of the units was 1 ft—in any case, the consignment was non-standard and the contractor had placed a definite order for 100 units. The contractor then realised that, in filling the joints between the units, he had used extra material because of the excessive width of the cracks, and thus, in effect, had himself provided an extra joist. To add insult to injury, the architect, seeing the spare joist lying around, required the contractor to remove it from the site at his expense. The contractor thus paid three times for the joist that he did not use.

Our study of the random variations of the properties of concrete is required to focus attention on the unwanted systematic variations which should be controlled.

H

# APPENDIX 1

## Mathematical Relationships

NORMAL PROBABILITY CURVE (see Chapter 1, page 29)

The equation of the normal probability curve[50] is given as

$$y = \frac{1}{\sigma\sqrt{2\pi}} \, e^{-\frac{(x-\bar{x})^2}{2\sigma^2}}$$

where y is the value of the ordinate giving unit area under the curve,

  $\bar{x}$ is the average value of the set of results of which x is any individual,

  $\sigma$ is the standard deviation of the results,

  e is the exponential function.

The corresponding values of x and y for constructing the curve with unit area below it are as follows:[64]

| ±x | 0 | ·2 | ·4 | ·6 | ·8 | 1·0 | 1·2 | 1·4 | 1·6 | 1·8 | 2·0 | 2·2 | 2·6 | 3·0 | 3·5 |
|---|---|---|---|---|---|---|---|---|---|---|---|---|---|---|---|
| y | ·399 | ·391 | ·368 | ·333 | ·290 | ·242 | ·194 | ·150 | ·111 | ·079 | ·054 | ·036 | ·014 | ·004 | ·001 |

CALCULATION OF TESTING ERROR FROM PAIRS OF RESULTS OBTAINED FROM INDIVIDUAL BATCHES (see Chapter 4, page 59)

The testing error may be calculated from a series of pairs of results taken from batches as follows:

Let $x_1$ and $x_2$ represent the pairs of results in any batch and m be the number of batches. Then the testing error

$$\sigma_t = \sqrt{\frac{\Sigma(x_1^2 + x_2^2) - \dfrac{\Sigma[(x_1 + x_2)^2]}{2}}{m}}$$

SIGNIFICANCE OF DIFFERENCE IN AVERAGE STRENGTHS OF TRIAL MIXES OF TWO TYPES OF CONCRETE (see Chapter 6, page 97)

The general procedure[65] for assessing the significance of the difference between the average values, $\bar{x}_1$ and $\bar{x}_2$, for two sets of results,

[111]

$x_1$, etc., and $x_2$, etc., containing $n_1$ and $n_2$ values, respectively, is as follows:

Calculate $\bar{x}_1 - \bar{x}_2$

$$\text{Calculate } \sigma = \sqrt{\frac{\Sigma(x_1 - \bar{x}_1)^2 + \Sigma(x_2 - \bar{x}_2)^2}{n_1 + n_2 - 2}}$$

$$\text{Calculate } t = \frac{\bar{x}_1 - \bar{x}_2}{\sigma} \sqrt{\frac{n_1 \times n_2}{n_1 + n_2}}$$

The value of t is then compared with values of t for $(n_1 + n_2 - 2)$ degrees of freedom for various chances of the difference $\bar{x}_1 - \bar{x}_2$ being fortuitous, tabulated in published works on statistics.[65]

However, in general, equal numbers of each type of trial mix will be made so that $n_1 = n_2$; the work can then be simplified, as follows:

$$\text{Calculate } \frac{\bar{x}_1 - \bar{x}_2}{\sqrt{\Sigma(x_1 - \bar{x}_1)^2 + \Sigma(x_2 - \bar{x}_2)^2}}$$

If this value is greater than the appropriate factor below, where n is the number of results for each type of concrete, the chance that the difference is not fortuitous is better than the percentage shown.

| n | 2 | 3 | 4 | 6 |
|---|---|---|---|---|
| 90 per cent. .. | 2·06 | 0·87 | 0·56 | 0·33 |
| 98 per cent. .. | 4·93 | 1·53 | 0·91 | 0·50 |

# APPENDIX 2

## Examples of Specification and Control Techniques

This appendix illustrates the use of the techniques described in the text with the help of two series of test cube results which have been obtained in practice; a very slight modification has been made to the general level of the strength in the first set, so that both represent borderline compliance with the requirements of assumed specifications.

### EXAMPLE OF NORMAL AND SUPPLEMENTARY SPECIFICATIONS

The data given in table 7 is used to illustrate the normal and supplementary methods of specifying concrete strength outlined in Chapter 4 (pages 66 to 69) and the use of control charts described in Chapter 5 (pages 76 to 82).

The outline specification assumed for this example is as follows:

(i) *Testing*

The quality of the hardened concrete shall be checked regularly by the following procedure:

1. The engineer shall select random batches of concrete for examination without warning the contractor.
2. Each batch selected shall be sampled on discharge from the mixer according to the method of Part 1 of B.S. 1881 : 1952.[2]
3. One 6-in cube shall be made, from the well-mixed sample, cured and tested at seven days according to the method of Parts 7 and 8 of B.S. 1881 : 1952.
4. At least six samples shall be taken on each full day's concreting for the first seven days or until at least 40 cubes have been made, whichever is the shorter; this is the initial period. Thereafter the number of cubes made per day may be reduced.

(ii) *Control*

The mix used shall be approved by the engineer in relation to the equipment and personnel provided by the contractor for producing the concrete as follows:

1. The facilities for controlling the quality of the concrete shall be at least as good as Standard of Control C of the I.C.E. Report.[46]
2. The standard deviation of cube results to be assumed in designing the mix shall be 850 lb/in$^2$ unless the contractor can produce

[113]

## TABLE 7
**Compressive strength results for examples of normal and supplementary methods of checking compliance with specification requirements**

| Day | Strength (lb/in²) | | | Day | Strength (lb/in²) | | | Day | Strength (lb/in²) | | |
|---|---|---|---|---|---|---|---|---|---|---|---|
| | Result | Aver. | Range | | Result | Aver. | Range | | Result | Aver. | Range |
| 1 | 4,950 | | | 8 contd. | 4,950 | | | 18 contd. | 3,550 | | |
| | 3,600 | | | | 2,700 | | | | 4,250 | | |
| | 3,800 | | | | 2,700 | | | 22 | 3,300 | | |
| | 3,750 | 4,025 | 1,350 | | 4,000 | 3,600 | 2,250 | | 3,300 | 3,600 | 950 |
| | 3,700 | | | | 3,400 | | | 23 | 3,550 | | |
| | 4,550 | | | | 3,300 | | | 24 | 3,500 | | |
| | 5,250 | | | | 3,800 | | | 25 | 4,000 | | |
| | 5,550 | 4,750 | 1,850 | | 3,200 | 3,425 | 600 | 29 | 2,300* | 3,350 | 1,700 |
| | 4,800 | | | End of normal method | | | | | 2,300* | | |
| 2 | 3,600 | | | 10 | 4,050 | | | 33 | 4,600 | | |
| | 4,200 | | | | 4,100 | | | 34 | 6,250 | | |
| | 3,600 | 4,050 | 1,200 | | 4,300 | | | 35 | 4,500 | 4,400 | 3,950 |
| | 3,800 | | | | 4,750 | 4,300 | 700 | 36 | 4,800 | | |
| | 4,200 | | | | 4,600 | | | 37 | 5,750 | | |
| | 4,700 | | | | 3,950 | | | 38 | 4,550 | | |
| | 4,700 | 4,350 | 900 | | 3,150 | | | 57 | 3,650 | 4,700 | 2,100 |
| | 4,850 | | | 12 | 3,550 | 3,800 | 1,450 | 59 | 5,050 | | |
| | 4,300 | | | | 3,450 | | | 63 | 3,100 | | |
| 3 | 4,350 | | | | 4,200 | | | 65 | 3,450 | | |
| | 4,100 | 4,400 | 750 | | 3,300 | | | 66 | 3,850 | 3,850 | 1,950 |
| | 3,950 | | | 13 | 3,500 | 3,600 | 900 | 69 | 4,950 | | |
| | 4,850 | | | | 3,550 | | | 70 | 4,750 | | |
| | 5,050 | | | | 4,600 | | | 83 | 3,100 | | |
| | 3,800 | 4,400 | 1,250 | | 5,500 | | | 96 | 3,050 | 3,950 | 1,900 |
| 6 | 5,700 | | | 14 | 4,350 | 4,500 | 1,950 | | 4,150 | | |
| | 4,100 | | | | 3,050 | | | 97 | 4,700 | | |
| | 3,800 | | | 15 | 3,050 | | | 98 | 2,850 | | |
| 7 | 3,300 | 4,225 | 2,400 | | 3,600 | | | 99 | 3,400 | 3,800 | 1,850 |
| | 3,700 | | | | 4,050 | 3,450 | 1,000 | 100 | 3,000 | | |
| | 3,550 | | | 16 | 2,500* | | | 102 | 3,850 | | |
| | 3,400 | | | | 3,900 | | | 103 | 3,350 | | |
| | 3,350 | 3,500 | 350 | | 3,600 | | | 104 | 3,450 | 3,400 | 850 |
| | 3,100 | | | 17 | 3,950 | 3,450 | 1,450 | 105 | 2,500* | | |
| 8 | 3,450 | | | | 3,300 | | | 106 | 3,350 | | |
| | 4,050 | | | | 4,250 | | | 107 | 3,950 | | |
| | 3,700 | 3,575 | 950 | 18 | 3,550 | | | | | | |
| | | | | | 3,800 | 3,725 | 950 | 115 | 5,600 | 3,850 | 3,100 |

First 44 results: Average—4,030 lb/in²; standard deviation—715 lb/in².
All results: Average—3,930 lb/in²; standard deviation—775 lb/in².
*Results at or below specified minimum.

satisfactory evidence that the means of controlling the concrete he is to provide will give a lower value, to be agreed by the engineer.

3. The mix shall be designed to have an average strength of 4,000 lb/in² unless the agreed value of the standard deviation, $\sigma$, is less than 850 lb/in²; in this case the design average strength will be 2,500 lb/in² $+ 1 \cdot 75\sigma$.

NOTE: This requirement corresponds to a minimum strength of 2,500 lb/in², below which 4 per cent. of the cube results may be expected to occur.

(iii) *Compliance*

The adequacy of the control of quality during the progress of the job shall be checked as follows:

1. Each consecutive set of four cube results shall be treated as a set and the average, $\bar{x}$, and the range, w, for each set shall be calculated.

2. If, during the initial period, more than one value of $\bar{x}$ occurs below $0 \cdot 64\sigma$ less than the design average, or if any value of $\bar{x}$ occurs below $0 \cdot 98\sigma$ less than the design average, the engineer will reserve the right to demand an approved change in mix proportions to increase the average strength.

3. Furthermore, if during the initial period any value of w exceeds $4\sigma$, the engineer will reserve the right to demand a change in mix proportions or control procedure or both.

4. At the end of the initial period the average, $\bar{x}_1$, and the standard deviation, $\sigma_1$, of the results obtained shall be calculated. If $\bar{x}_1 - 1 \cdot 75\sigma_1$ exceeds 2,850 lb/in², the design average strength may be reduced by an amount to be agreed between the engineer and the contractor. Compliance with the requirements shall then be checked against the appropriate limits for the new mix design criteria.

NOTE: The value of 2,850 lb/in² ($=2,500 + 350$) is based on table 2: for $\sigma = 600$ lb/in² the limit would be $+ 250$ (by interpolation); hence for $\sigma = 850$ lb/in² the limit is $+ 350$.

5. If, after the initial period, there is in any 10 consecutive sets
   (a) more than one value of $\bar{x}$ below $0 \cdot 64\sigma$ less than the design average strength
   (b) any value of $\bar{x}$ below $0 \cdot 98\sigma$ less than the design average strength
   or (c) any value of w above $4\sigma$

the average of the N results obtained up to that time shall be determined and if found to be below $1 \cdot 96\sigma \div \sqrt{N}$ less than the design average strength the engineer will reserve the right to demand an approved change in the mix proportions.

In considering the example it is assumed that the standard deviation was taken to be 850 lb/in². The initial period was taken to the last full set of four results on day 8 and included 11 sets.

The control levels are:

$\bar{x}$ : $\pm 0 \cdot 64 \times 850$     and     $\pm 0 \cdot 98 \times 850$
$= \pm 545$ lb/in²          $= \pm 830$ lb/in²

Average w:    $2 \cdot 06 \times 850 = 1{,}750$ lb/in²
Maximum w:    $4 \cdot 0 \times 850 = 3{,}400$ lb/in²
Minimum w:   $0 \cdot 59 \times 850 = \phantom{3{,}}500$ lb/in²

The control charts are shown in figs. 13 and 14 (page 80).

It will be seen from the control chart for the average, $\bar{x}$, (fig. 13) that during the initial period the average strength was adequate or perhaps a little higher than necessary, and from fig. 14 that the variation was also better than necessary.

Calculation of $\bar{x}_1 - 1 \cdot 75\sigma_1$ showed this to be:

$$4{,}030 - 1 \cdot 75 \times 715 \text{ lb/in}^2$$
$$= 4{,}030 - 1{,}250 = 2{,}780 \text{ lb/in}^2$$

There was therefore neither opportunity nor point in amending the design of the mix, but the job had got off to a good start.

After the initial period the average strength tended to drop, but the occurrences of values of $\bar{x}$ outside the limits were so marginal that it is assumed that no adjustment to the mix was demanded, especially since the values of w were low.

The overall calculated minimum strength was:

$$3{,}930 - 1 \cdot 75 \times 775$$
$$= 3{,}930 - 1{,}360$$
$$= 2{,}570 \text{ lb/in}^2$$

The R.P.I. (see page 82) for the initial period

$$= \frac{4{,}000 - 2{,}500}{\frac{1}{2} \times 1{,}260} = 2 \cdot 4$$

and for the whole job

$$= \frac{1{,}500}{\frac{1}{2} \times 1{,}500} = 2 \cdot 0$$

The anticipated R.P.I. was:

$$\frac{1,500}{\frac{1}{2} \times 1,750} = 1 \cdot 7$$

An R.P.I. of $3 \cdot 0$ represents $0 \cdot 1$ per cent. of the results below the specified minimum.

## EXAMPLE OF CONTROL CHARTS FOR STANDARD DEVIATION AND COEFFICIENT OF DISPLACEMENT

The properties of sets of 10 results have been recorded in table 8 for a job where 410 test cubes were made; the details of an assumed specification are also given.

Fig. 19 shows control charts for values of the average, $\bar{x}$, the actual standard deviation, s, the calculated minimum value, $\bar{x} - 1 \cdot 28\sigma$, and the coefficient of displacement, u. In addition, histograms of these properties have been drawn to show their distribution. It will be seen that the distribution of the average is approximately normal, even though there is a slight tendency for the level to drop with time; but the distributions of the standard deviation, calculated minimum value and coefficient of displacement tend to be skew.

The control levels for enclosing 95 per cent. of the values (for sets of 10 results) are given as follows:

$\bar{x}$: $\pm 0 \cdot 620 \times 800 = \pm 495$ lb/in$^2$
s: $0 \cdot 520 \times 800 = 415$ lb/in$^2$
and $1 \cdot 379 \times 800 = 1,100$ lb/in$^2$
$\bar{x} - 1 \cdot 28\sigma$: approx. $\pm 700$ lb/in$^2$ (Ref. 48)
u: $0 \cdot 62$ and $2 \cdot 80$

and the average values are:

$\bar{x}$: $6,500$ lb/in$^2$
s: $0 \cdot 923 \times 800 = 740$ lb/in$^2$
$\bar{x} - 1 \cdot 28\sigma$: $5,500$ lb/in$^2$
u: $1 \cdot 28$

The overall values for the job are:

$\bar{x}$: $6,470$ lb/in$^2$
$\sigma$: $795$ lb/in$^2$
$\bar{x} - 1 \cdot 28\sigma$: $5,450$ lb/in$^2$
$$\frac{\bar{x} - 5,500}{\sigma} = 1 \cdot 22$$

## TABLE 8

### Summary of 410 results illustrating control chart techniques for a large job
(see fig. 19)

The table gives properties of sets of 10 compressive strength results rounded off to the nearest 10 lb/in².

$\bar{x}$ = average for set; s = actual standard deviation of set

$\bar{x} - 1 \cdot 28\sigma$ = calculated minimum value; $u = \dfrac{\bar{x} - L}{s}$ = coefficient of displacement, where L = specified minimum strength

Overall average = 6,470 lb/in²; overall standard deviation = 795 lb/in².

It is assumed that the specification required a minimum value, L, = 5,500 lb/in², with a probability that 10% of the results would fall below this value, and that the standard deviation to be expected was 800 lb/in². The design average strength would therefore be 5,500 + 1·28 × 800 lb/in² = 6,500 lb/in².

| Set no. | Strength (lb/in²) | | | u | Set no. | Strength (lb/in²) | | | u |
|---|---|---|---|---|---|---|---|---|---|
| | $\bar{x}$ | s | $\bar{x}-1\cdot28\sigma$ | | | $\bar{x}$ | s | $\bar{x}-1\cdot28\sigma$ | |
| 1 | 6,260 | 1,040 | 4,860 | 0·73 | 21 | 6,770 | 740 | 5,780 | 1·72 |
| 2 | 6,670 | 470 | 6,040 | 2·50 | 22 | 5,880 | 540 | 5,160 | 0·70 |
| 3 | 6,710 | 550 | 5,970 | 2·20 | 23 | 5,980 | 680 | 5,060 | 0·71 |
| 4 | 6,350 | 980 | 5,030 | 0·87 | 24 | 6,340 | 1,250 | 4,650 | 0·67 |
| 5 | 6,860 | 690 | 5,920 | 1·97 | 25 | 6,400 | 450 | 5,800 | 2·00 |
| 6 | 7,290 | 540 | 6,560 | 3·31 | 26 | 6,660 | 500 | 5,990 | 2·32 |
| 7 | 7,230 | 350 | 6,760 | 4·95 | 27 | 7,190 | 470 | 6,560 | 3·60 |
| 8 | 6,580 | 860 | 5,420 | 1·26 | 28 | 6,260 | 730 | 5,280 | 1·04 |
| 9 | 6,920 | 740 | 5,920 | 1·92 | 29 | 6,450 | 680 | 5,530 | 1·40 |
| 10 | 6,690 | 400 | 6,150 | 2·97 | 30 | 5,960 | 920 | 4,720 | 0·50 |
| 11 | 6,360 | 520 | 5,660 | 1·66 | 31 | 6,280 | 930 | 5,030 | 0·84 |
| 12 | 6,660 | 730 | 5,670 | 1·59 | 32 | 6,820 | 290 | 6,440 | 4·55 |
| 13 | 6,360 | 520 | 5,650 | 1·66 | 33 | 6,740 | 300 | 6,330 | 4·14 |
| 14 | 6,670 | 590 | 5,870 | 1·98 | 34 | 6,570 | 420 | 6,010 | 2·55 |
| 15 | 7,060 | 520 | 6,360 | 3·00 | 35 | 6,300 | 610 | 5,480 | 1·31 |
| 16 | 6,350 | 480 | 5,710 | 1·77 | 36 | 5,800 | 750 | 4,790 | 0·40 |
| 17 | 5,160 | 710 | 4,200 | -0·48 | 37 | 6,320 | 760 | 5,290 | 1·08 |
| 18 | 6,700 | 640 | 5,850 | 1·88 | 38 | 6,580 | 690 | 5,640 | 1·53 |
| 19 | 6,240 | 560 | 5,480 | 1·32 | 39 | 6,020 | 780 | 4,980 | 0·67 |
| 20 | 6,800 | 910 | 5,570 | 1·43 | 40 | 6,010 | 590 | 5,210 | 0·86 |
| | | | | | 41 | 5,910 | 530 | 5,200 | 0·77 |
| Average | | | | | | 6,470 | 640 | 5,610 | |

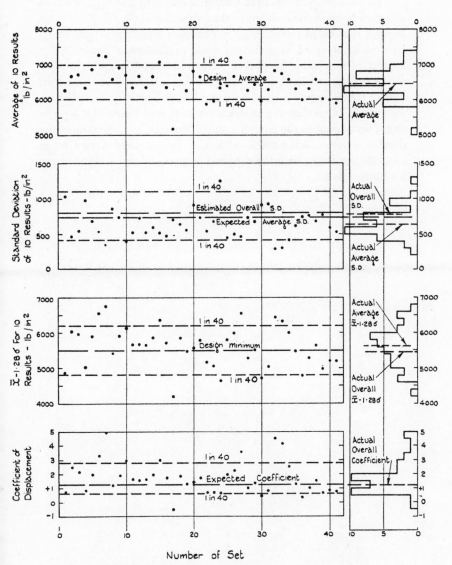

Number of Set

**Fig. 19—Comparison of control charts for a large job**

As has been pointed out in Chapter 4 (pages 64 and 65), the values of s tend to be appreciably lower than $\sigma$ for the whole job if it is large because the sets of 10 results do not embrace long-term trends which have to be accepted as occurring while production is nominally 'in control'. On the other hand, these same trends result in a wider scatter of the average values, $\bar{x}$, with the result that a higher proportion than 1 in 40 occurs beyond each of the control levels—both upper and lower. As a result, more than the expected proportion of values of $\bar{x} - 1\cdot28\sigma$ and u occur outside the control levels. However, in this example the number of values below the lower levels—the critical ones—are not as seriously out of line with the expected proportions as the values above the upper levels.

# APPENDIX 3

## References and Bibliographical Reviews

REFERENCES

1. BRITISH STANDARDS INSTITUTION: 'Methods for the sampling and testing of mineral aggregates, sands and fillers'. London, pp. 84, B.S. 812 : 1960.
2. BRITISH STANDARDS INSTITUTION: 'Methods of testing concrete'. London, pp. 61, B.S. 1881 : 1952.
3. RICHARDSON, J. T.: 'The reduction and presentation of experimental results'. London, pp. 43, B.S. 2846 : 1957.
4. AKROYD, T. N. W. (1961): 'The accelerated curing of concrete test cubes', *Proceedings of the Institution of Civil Engineers*, **19**, 1.
   (1962): Discussion, **21**, 678.
   (For earlier references on this subject, see list of references at the end of this paper and the discussion.)
5. GRAHAM, G. and MARTIN, F. R. (1946): 'Heathrow. The construction of high-grade quality concrete paving for modern transport aircraft', *Journal of the Institution of Civil Engineers*, **26**, 117. Discussion, p. 191.
6. Sparkes, F. N. (1949): 'Control of variations in quality of concrete and its effect on mix proportions', *The Reinforced Concrete Review*, **1**, 543.
7. MURDOCK, L. J. (1953): 'The control of concrete quality', *Proceedings of the Institution of Civil Engineers, Part I*, **2**, 426.
   (1954): Discussion, Part I, **3**, 233.
8. HIMSWORTH, F. R. (1954): 'The variability of concrete and its effect on mix design', *Proceedings of the Institution of Civil Engineers, Part I*, **3**, 163.
   (1954): Discussion, **3**, 736.
9. McINTOSH, J. D. (1955): 'Basic principles of concrete mix design'. Cement and Concrete Association. Proceedings of a symposium on mix design and quality control of concrete. London, May 1954, p. 3.
10. SPARKES, F. N. (1955): 'The control of concrete quality: a review of the present position'. Cement and Concrete Association. Proceedings of a symposium on mix design and quality control of concrete. London, May 1954, p. 211.
11. THOMAS, F. G. (1955): 'Quality control and its effect on structural design'. Cement and Concrete Association. Proceedings of a symposium on mix design and quality control of concrete. London, May 1954, p. 283.
12. FISHER, J. M. (1955): 'Quality control for road and airfield construction'. Cement and Concrete Association. Proceedings of a symposium on mix design and quality control of concrete. London, May 1954, p. 345.
13. DICK, W. (1955): 'Quality control for mass concrete with special reference to dams and related hydraulic structures'. Cement and Concrete Association. Proceedings of a symposium on mix design and quality control of concrete. London, May 1954, p. 369.

14.  MERCER, L. B. (1955): 'Ready-mixed concrete: quality control refinements'. Cement and Concrete Association. Proceedings of a symposium on mix design and quality control of concrete. London, May 1954, p. 409.

15.  HIMSWORTH, F. R. (1955): 'The application of statistics to concrete quality'. Cement and Concrete Association. Proceedings of a symposium on mix design and quality control of concrete. London, May 1954, p. 465.

16.  COLLINS, A. R. (1955): 'Methods of specifying concrete'. Cement and Concrete Association. Proceedings of a symposium on mix design and quality control of concrete. London, May 1954, p. 488.

17.  ABDUN-NUR, E. A. (1962): 'How good is good enough ?', *Proceedings of the American Concrete Institute*, **59**, 31.

18.  BRITISH STANDARDS INSTITUTION: 'The structural use of normal reinforced concrete in buildings'. London, pp. 83, CP 114 (1957).

19.  DICK, W.: Contribution to discussion of Ref. 9, p. 23.

20.  ABDUN-NUR, E. A., and WADDELL, J. J. (1959): 'Control of concrete mixes', *Journal of the American Concrete Institute*, **30**, No. 9. (*Proceedings*, **55**), 947.
     (1959): Discussion, **31**, No. 3. (*Proceedings*, **56**), 1543.

21.  DAY, K. W. (1961): 'The specification of concrete', *Constructional Review*, **34**, 44.

22.  The Engineering Index, Engineering Index, Inc., New York. Volumes for 1945 to 1954, inclusive: See under heading 'Structural design—safety factor'.

23.  PUGSLEY, A. G. (1951): 'Concept of safety in structural engineering', *Journal of the Institution of Civil Engineers*, **36**, 5.

24.  LEVY, H. (1953): 'The impact of statistics on civil engineering', *Proceedings of the Institution of Civil Engineers, Part I*, **2**, 681.

25.  BAKER, A. L. L. (1958): 'The work of the European Committee for concrete', *The Structural Engineer*, **36**, 10.
     (1958): Discussion, **36**, 271.
     (1958): *Reinforced Concrete Review*, **4**, 637.

26.  BAKER, A. L. L. (1959): 'European Committee on concrete: Report of a meeting at Vienna in April, 1959', *Reinforced Concrete Review*, **5**, 247.
     (1960): Discussion, **5**, 401.

27.  MATTHEWS, D. D. (1958): 'Load-factor design in building regulations: future British practice'. Cement and Concrete Association. Proceedings of a symposium on the strength of concrete structures. London, May 1956, p. 626.

28.  INSTITUTION OF STRUCTURAL ENGINEERS (1955): 'Report on structural safety', *The Structural Engineer*, **33**, 141.
     (1956): Discussion, **34**, 307.

29.  HORNE, M. R. (1958): 'Some results of the theory of probability in the estimation of design loads'. Cement and Concrete Association. Proceedings of a symposium on the strength of concrete structures. London, May 1956, p. 3.

30. JOHNSON, A. I. (1958): 'The determination of the design factor for rein-forced concrete structures'. Cement and Concrete Association. Proceedings of a symposium on the strength of concrete structures. London, May 1956, p. 25.

31. PUGSLEY, A. G. (1958): 'Current trends in the specification of structural safety'. Cement and Concrete Association. Proceedings of a symposium on the strength of concrete structures. London, May 1956, p. 49.

32. GOLDSTEIN, A.: Contribution to discussion of Ref. 11, p. 292.

33. ASPLUND, S. O. (1958): 'The risk of failure', The Structural Engineer, 36, 268.

34. VORLICEK, M.: Contribution to discussion of Ref. 59.

35. MCINTOSH, J. D.: Contribution to discussion of Ref. 11, p. 290.

36. D.S.I.R. (BUILDING RESEARCH STATION) (1959): 'Principles of modern building, Vol. 1'. London, H.M.S.O., 3rd ed., pp. 4-6.

37. PLUM, N. M. (1953): 'Quality control of concrete—its rational basis and economic aspects', Proceedings of the Institution of Civil Engineers, Part I, 2, 311.
(1954): Discussion, 3, 99.

38. RUDJORD, A. (1960): 'Formulering av fasthetskrav til betong' ('Formulation of requirements for concrete'), Teknisk Ukeblad, No. 38, 849. (Summary in English).

39. THE INSTITUTION OF CIVIL ENGINEERS RESEARCH COMMITTEE (1962): 'Ultimate load design of concrete structures', Proceedings of the Institution of Civil Engineers, 21, 399.

40. BRITISH STANDARDS INSTITUTION: 'The structural use of prestressed concrete in buildings'. London, pp. 44, CP 115 : 1959.

41. MCINTOSH, J. D.: Contribution to discussion of Ref. 30, p. 61.

42. BRITISH STANDARDS INSTITUTION: 'Design and construction of reinforced and prestressed concrete structures for the storage of water and other aqueous liquids'. London, pp. 50, CP 2007 (1960).

43. ERNTROY, H. C. (1960): 'The variation of works test cubes'. Cement and Concrete Association. Research Report No. 10, pp. 28.

44. ACI COMMITTEE 214 (1957): 'Recommended practice for evaluation of compression test results of field concrete', Journal of the American Concrete Institute, 29, No. 1. (Proceedings, 54), 1.
(1958): Discussion, 29, No. 9. (Proceedings, 54), 775.

45. ABDUN-NUR, E. A., and TUTHILL, L. H. (1959): 'Criteria for modern specifications and control', Journal of the American Concrete Institute, 30, No. 7. (Proceedings, 55), 759.
(1959): Discussion, 31, No. 3. (Proceedings, 56), 1479.

46. INSTITUTION OF CIVIL ENGINEERS RESEARCH COMMITTEE (1955): 'Quality of concrete in the field', Proceedings of the Institution of Civil Engineers, Part I, 4, 336.

47. DUDDING, B. P., and JENNETT, W. J.: 'Quality control charts'. London, British Standards Institution, pp. 89, B.S. 600R : 1942 (withdrawn).

48. GUSTAFFSON, S. (1961): 'Statistisk bedömning av betongkvaliteten' ('Statistical estimation of concrete quality'), *Nordisk Betong*. **5**, 305. (English summary).

49. DUDDING, B. P., and JENNETT, W. J.: 'Control chart technique when manufacturing to a specification'. London, British Standards Institution, pp. 77, B.S. 2564 : 1955.

50. PEARSON, E. S.: 'The application of statistical methods to industrial standardisation and quality control'. London, British Standards Institution, (a) pp. 161, B.S. 600 : 1935. (b) Reprinted 1960 (with minor amendments), pp. 152, B.S. 600 : 1935.

51. ERNTROY, H. C.: Contribution to discussion of Ref. 7, p. 236.

52. BRITISH STANDARDS INSTITUTION: 'Portland cement (ordinary and rapid-hardening)'. London, pp. 37, B.S. 12 : 1958.

53. SNOW, F. S., LEA, F. M., MURDOCK, L. J., and BURKE, E. (1953): 'Would the strength grading of ordinary Portland cement be a contribution to structural economy?', *Proceedings of the Institution of Civil Engineers, Part III*, **2**, 448.

54. BRITISH STANDARDS INSTITUTION: 'Concrete aggregates from natural sources'. London, pp. 16, B.S. 882, 1201 : 1954.

55. KIERKEGAARD-HANSEN, P. (1961): 'Bemaerkninger til en ofte anvendt prøvemetode' ('Remarks on a frequently used testing method'), *Nordisk Betong*, **5**, 391. (English summary).

56. KAPLAN, M. F. (1958): 'Compressive strength and ultrasonic pulse velocity relationships for concrete in columns', *Journal of the American Concrete Institute*, **29**, No. 8. (*Proceedings*, **54**), 675. (1958): Discussion, **30**, No. 3. (*Proceedings*, **54**), 1259.

57. KOLEK, J. (1958): 'An appreciation of the Schmidt rebound hammer', *Magazine of Concrete Research*, **10**, 27. (1958): Discussion, **10**, 144.

58. MCINTOSH, J. D., and MURPHY, W. E.: Contribution to discussion of Ref. 56.

59. MASTERMAN, O. J. (1958): 'Quality control in prestressed concrete production', *Magazine of Concrete Research*, **10**, 57. (1959): Discussion, **11**, 39.

60. BRITISH STANDARDS INSTITUTION: 'Concrete bricks and fixing bricks'. London, pp. 12, B.S. 1180 : 1944.

61. BRITISH STANDARDS INSTITUTION: 'Precast concrete blocks'. London, pp. 19, B.S. 2028 : 1953.

62. BRITISH STANDARDS INSTITUTION: 'Structural recommendations for load-bearing walls'. London, pp. 47, CP 111 (1948).

63. BRITISH STANDARDS INSTITUTION: 'Sandlime (calcium silicate) bricks'. London, pp. 14, B.S. 187 : 1955.

64. WRIGHT, P. J. F. (1954): 'Statistical methods in concrete research', *Magazine of Concrete Research*, **5**, 139.

65. BROWNLEE, K. A. (1949): 'Industrial experimentation'. London, H.M.S.O., 4th Ed., pp. 194. (Reprinted 1957).

ADDITIONAL REFERENCES

### General works on statistics

66. A.S.T.M. COMMITTEE E-11 (1951): 'A.S.T.M. manual on quality control of materials'. Philadelphia, American Society for Testing Materials, pp. 127. Special Technical Publication 15-C.
67. HALD, A. (1952): 'Statistical theory with engineering applications'. Wiley, New York (Chapman and Hall, London), pp. 783.
68. MORONEY, M. J. (1956): 'Facts from figures'. Harmondsworth, Penguin Books Ltd., 3rd Ed., pp. 472. Pelican Books A236.

### Use of statistics in concrete technology

69. CORDON, W. A. (1956): 'Size and number of samples and statistical considerations in sampling'. Philadelphia, American Society for Testing Materials. Significance of tests and properties of concrete and concrete aggregates, p. 14. A.S.T.M. Special Technical Publication No. 169.
70. O'LOUGHLIN, D. J. (1960): 'Concrete strength specification in accordance with S.A.A. Code CA2—1958', Constructional Review, 33, 23.
71. PRATT, H. A. (1956): 'Evaluation of test results'. Philadelphia, American Society for Testing Materials. Significance of tests and properties of concrete and concrete aggregates, p. 22. A.S.T.M. Special Technical Publication No. 169.
72. WARRIS, B. (1961): 'Uttagning av prov och statistisk behandling och värdering av provningsresultaten' ('Sampling as related to statistical treatment and evaluation of test results'), Nordisk Betong, 5, 179. (Includes valuable bibliography of Scandinavian papers on statistical methods.)

### Concrete research investigations involving statistical analysis

73. DAVIES, R. D. (1951): 'Some experiments on the compaction of concrete by vibration', Magazine of Concrete Research, 3, 71.
74. NEVILLE, A. M. (1956): 'The influence of size of concrete test cubes on mean strength and standard deviation', Magazine of Concrete Research, 8, 101.
   (1957): Discussion, 9, 52.
75. NEVILLE, A. M. (1959): 'The relation between standard deviation and mean strength of concrete test cubes', Magazine of Concrete Research, 11, 75.
   (1960): Discussion, 12, 44.
76. WILLIAMS, T. E. H. (1952): 'Vibrated concrete and mortar cubes: distribution of compressive strength', Magazine of Concrete Research, 3, 107.
77. WRIGHT, P. J. F. (1952): 'The effect of the method of test on the flexural strength of concrete (with an appendix on the statistical aspect by F. Garwood)', Magazine of Concrete Research, 4, 67.
78. WRIGHT, P. J. F. (1958): 'Variations in the strength of Portland cement', Magazine of Concrete Research, 10, 123.
   (1959): Discussion, 11, 104.

I

## REVIEW OF MIX DESIGN AND QUALITY CONTROL

5. Included a statistical study of the variation in strength of cement test cubes and concrete cubes, and suggested coefficients of variation appropriate to different types of control.

6. Indicated how improved control led to a more economical mix when designed to have a specified minimum strength; related variability with a particular type of control to the coefficient of variation.

7. Included a statistical analysis of concrete cube results from several sites and suggested limiting cube strengths for different types of control based on constant standard deviation for average strengths above 3,000 lb/in² and constant coefficient of variation below 3,000 lb/in².

8. The first part dealt with statistical principles used in mix design and concluded that variability for a particular type of control is related to standard deviation; the second part dealt with the effect of cement quality, cement/water ratio and testing error on concrete strength.

9. Included an assessment of the statistical implication of the specified minimum strength as a routine stage in the design of mixes.

10. Indicated the economy of designing a mix on the basis of a known variability of control with improved control techniques, and suggested design data; an appendix dealt with methods of measuring variation.

11. Suggested how structures can be designed more economically for a given degree of safety and durability if variation of concrete quality is controlled.

12. Indicated effects of variation in cement and aggregates on variation in concrete strength; suggested the use of control charts in maintaining control.

13. Considered effect of degree of control on economy of mix proportions; discussed statistical implications of concrete test results in relation to specifications and described use of control charts for averages of sets of results.

14. Included details of methods of applying statistical analysis to control of ready-mixed concrete, the degree of control being related to the coefficient of variation.

15. Dealt with the theory of statistics and its application to the significance of test results; mathematical definitions and procedures were given in an appendix.

16. Dealt with the problems of specifying concrete strength; suggested specification of strength in terms of a calculated statistical minimum value.


## REVIEW OF STRUCTURAL SAFETY

22. These volumes of the 'Engineering Index' list papers by A. M. Freudenthal, M. Prot, R. Levi, P. Weidlinger and A. Defay (and their discussion) and give summaries of their relevance to statistical aspects of structural safety.

23. A historical review of concepts of safety in a wide range of structures and for a variety of materials.

24. A mathematician's approach to some engineering problems. A considerable part of the paper is given to a particular example involving concrete pillars. For comment by an engineer, see reference 32.

25. Reviews research data presented to the Committee, including histograms of ratio of observed value to calculated value of critical characteristics of failure for large numbers of tests, and accepts a statistical formula for determining factors of safety. The discussion includes further comment on choice of load factors.

26. Records further experimental data presented to the Committee in statistical form. The discussion includes references to the safety factor for over- and under-reinforced beams and to the statistical approach to design.

27. Discusses the practical application of the load-factor method to the design of reinforced concrete sections as proposed for the British Standard Code of Practice CP 114.

28. Deals with two categories of probability of collapse; the first includes workmanship, loading and accuracy of analysis, and the second, danger to personnel and economic consideration. Four appendices deal with different ways of applying the concepts in practice. The discussion dealt largely with the application of the report to design practice.

29. A study of the statistical distribution of loadings to which structures are subjected; includes also a statistical examination of 'shake-down' analysis.

30. Analyses the causes of variation in ultimate strength of a particular design of floor slab, and the loads to which it will be subjected from people and furniture; these are related to an overall economic design.

31. Deals with five trends: the importance of the ultimate strength of the structure; the value of statistics and probability theory; the development of the idea of a design life for a structure; the linking of safety with economic effects of failure; interest in the specification of safety from a human risk standpoint.

33. Discusses the statistical approach to safety and, in particular, the principle of least total mortality.

39. Discusses design criteria for frames and slabs, including alternative recommendations for choice of load factors.


## REVIEW OF GENERAL WORKS ON STATISTICS

3. A guide to establishing the significance of test results, with examples related to a set of 400 concrete cube results.

47. Although now withdrawn, gives data for establishing confidence limits and control limits for 'coefficient of displacement' as well as average, range and standard deviation of sets of results.

49. Superseded B.S. 600R (Ref. 47): gives data for producing control charts; mainly related to control of dimensional tolerances.

50. A general review of the use of statistics in establishing specification limits and controlling quality to comply with the specified requirements.

65. Deals with statistical procedures applicable to research investigations.

66.   Deals with statistical theory, presentation of data, uncertainty of estimates of characteristics of the consignment and control charts.
67.   A mathematical treatise on statistics with some examples drawn from cement technology.
68.   A general introduction to statistics for the layman.

# Index

1. *Items are indexed by paragraphs starting on the pages given.*
2. *Some entries refer to the ideas expressed by the item and not necessarily the actual words listed.*
3. *An asterisk (\*) indicates that the item occurs in more than one paragraph on the page.*
4. *(t) or (f) after a page number indicates a table or figure, respectively, where it does not occupy a whole page.*